Aerofoils for Aeromodellers

Martyn Pressnell

Pitman

First Published 1977

Pitman Publishing Ltd
Pitman House, 39 Parker Street, London WC2B 5PB

Pitman Medical Publishing Co Ltd
42 Camden Road, Tunbridge Wells, Kent TN1 2QD

Focal Press Ltd
31 Fitzroy Square, London W1P 6BH

Pitman Publishing Corporation
Fearon Publishers Inc
6 Davis Drive, Belmont, California 94002, USA

Pitman Publishing Pty Ltd
Pitman House, 158 Bouverie Street, Carlton, Victoria 3053, Australia

Pitman Publishing
Copp Clark Publishing
517 Wellington Street West, Toronto M5V 1G1, Canada

Sir Isaac Pitman & Sons Ltd
Banda Street, PO Box 46038, Nairobi, Kenya

Pitman Publishing Co SA (Pty) Ltd
Craighall Mews, Jan Smuts Avenue, Craighall Park,
Johannesburg 2001, South Africa

ISBN 0 273 01015 8

Text set in 11/12 pt IBM Press Roman, printed by photolithography, and bound in Great Britain at The Pitman Press, Bath

G.26:14

Contents

Acknowledgements

I would like to express my gratitude to the Hatfield Polytechnic for the use of its DEC System 10 computer facility, without which this book could not have been produced.

My grateful thanks are also due to Ron Moulton, Managing Editor of Model and Allied Publications Ltd, for his kind advice and for the provision of most of the photographs.

Illustrations

An open class rubber driven model, designed, constructed and displayed here by the author Martyn Pressnell. It uses NACA 6409 wing section with a forward spar to act as flow turbulator. *Photograph: Bob Frost.*

Foreword

All creative aeromodellers are aerofoil doodlers and thousands of successful models are a testimony to the general efficiency of the back of envelope 'zip-zip' curves which pass for rib sections. Blessed with an inherent eye for what looks near-right, the intelligent enthusiast finds the search for improved performance to be one of the greatest challenges in modelling. The aerofoil section plays a very great part in this demand for better flight characteristics and yet, surprisingly, it remains one of the least understood aspects of model aerodynamics.

Much of this situation is due to lack of information. There have been excellent reports by the Low Speed Aerodynamics Research Association, individual contributions to magazines and of course the well known NACA papers, but progress has been erratic and there is little for the academic reader to study.

Moreover, for Mr average-modeller, the scant information on aerofoils has largely been in the nature of one particular chord length as an illustration, supported by the tables of co-ordinates from which the outline may be plotted, and only occasionally with a handy polar diagram to indicate performance at useful Reynold's Numbers.

The subject is also provocative in that whenever any student of the subject has endeavoured to express his theories to explain causes for well-known and accepted results, he has been taken to task on points of detail. This situation has persisted for over thirty years and stems largely from the fact that the published works have either been over specialised or over simplified. After all – what *does* the average model designer want to know?

He needs to have actual size profiles, ready to employ. He requires advice on which is suitable for what category of model, and he needs sound information on the basics of model aerodynamics.

This book sets out to fulfil those requirements and adds a new approach to technical information for aeromodellers which will be quickly recognised as a milestone of progress. No longer is it good enough to settle for the 'straight' aerofoil co-ordinates. For efficiency one has to consider wing taper ratios, and span-wise section changes. The rib becomes a core, within the true aerofoil so that sections have to be skimmed to allow for sheeting. Above all, the sections have to be very carefully selected thoroughbreds which promise highest performance in a demanding hobby.

Martyn Pressnell is particularly well qualified to produce the answers to these demanding specifications. A practising aeromodeller since school-days, designer of very popular plan models, a regular competitor with numerous successes in SMAE events and club rallies, Martyn also has a unique academic background.

For over twelve years he was with the famous Handley Page design office, much involved in the creation of the *Jetstream*. In spare moments he founded the Hertfordshire Pedal Aeronauts, and led this group which made, and flew the world's first two-man-powered aircraft to fly – the *Toucan* which in itself could be called the largest ever model aeroplane! It is 139 ft span, yet it flies at little over 20 mph.

Currently Martyn is a Senior Lecturer in Aeronautical Engineering at The Hatfield Polytechnic, where low speed tunnels and computor technology offer finest opportunities for advancement of new concepts in aeronautics.

In his contribution to the technical reference books for students of low-speed flight, he provides an admirable blend of useful data on aerodynamics with extremely practical computer-drawn aerofoils, which will do much to improve understanding and model performance.

Ron Moulton

General Introduction

It is almost certainly true that more model aircraft are designed and constructed than all other types of flying machine. Correspondingly the variety of aerofoil sections in use for model wings is very extensive, and there is always a demand by modellers for geometric data and other information about sections which have proved successful for aeromodelling.

The types of section which have proved most suitable over the years are those developed and tested in the early 1930s by the National Advisory Committee for Aeronautics (NACA) in the USA, and at Göttingen in Germany. But inevitably as full-size aerodynamics has marched on into the realms of high speed flight, the needs of aeromodelling have been left behind. Fortunately a few people with the necessary scientific knowledge, aeromodelling interest, and access to wind tunnels, have continued with this pioneering work, notable amongst whom are George Benedek of Hungary, and Max Hacklinger and Dr R. Eppler of West Germany.

An important contribution in its time was made by F. W. Schmitz whose book *Aerodynamik des Flugmodells* was published in Germany in 1939. This was translated and made available in Great Britain by the Ministry of Aircraft Production shortly afterwards; it brought to the attention of modellers and others the effects of surface roughness and turbulent flow on aerofoil characteristics at very slow flying speeds. Soon after the Second World War the Low Speed Aerodynamic Research Association (LSARA) and the Model Aircraft Research Project (MARP) produced some useful and interesting information.

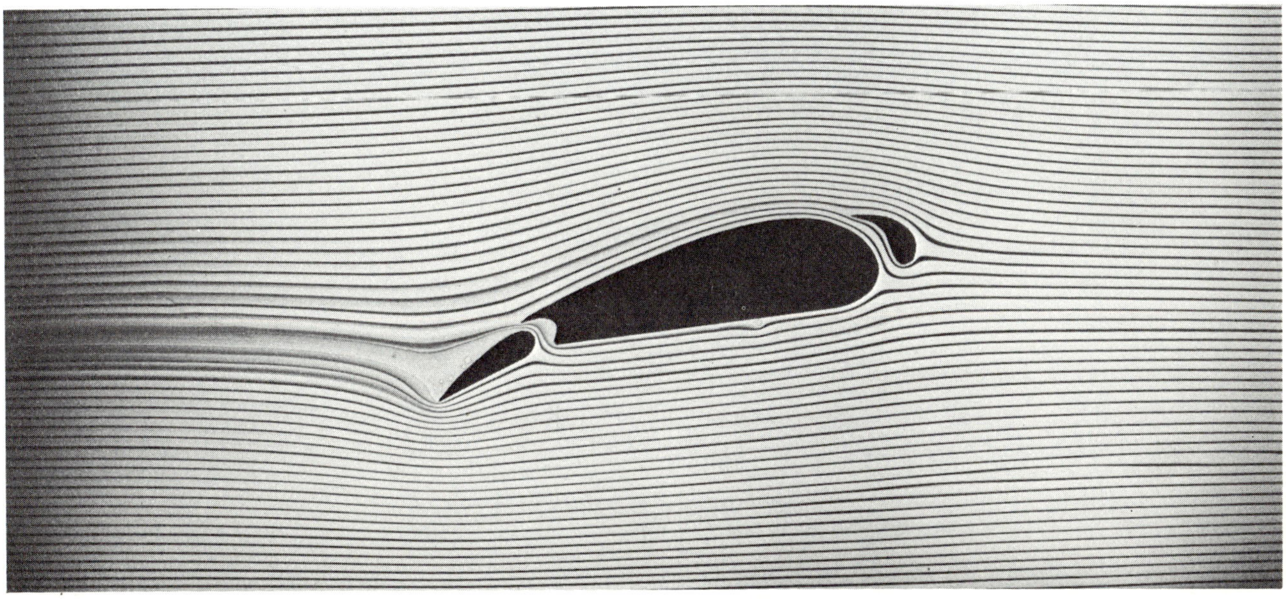

STREAMLINE FLOW

Visualisation of the flow pattern around an aerofoil with trailing edge flap and leading edge slat, as obtained by introducing filaments of dye into water passing the section. The variation in thickness of the paths between the streamlines is clearly visible. *By kind permission of the Kingston Polytechnic.*

However, since those times the scientific development of low speed aerofoils has been very slow and spasmodic in the UK.

It is in the nature of aeromodelling that development of all aspects of model flight continues apace on a trial and error basis and this is quite effective because of the large numbers of models constructed, and flown competitively. The model press throughout the world disseminates the results by publishing details of the most successful models and the regeneration of thoroughbred models continues. The performance of aerofoil sections is influenced by the nature of the wing construction, which affects both the actual shape and surface finish. Not surprisingly, ingenious modellers have taken advantage of this in 'developing' their own aerofoils. The sections are usually named after their originators, or referred to by the name of the model on which they were used.

However, methods of trial and error development are not entirely satisfactory, usually revealing 'effect' but not 'cause', in what may in fact be extremely small improvements in performance. In view of the large numbers of models built and tested, some sort of statistical appraisal may be possible. Allnutt and Kaczanowski in Canada have studied A2 glider performance on this basis, which is reported in *Aeromodeller Annual 1970–71*. Further testing of model aerofoils is long overdue, and we must hope that more work of this sort will eventually be possible.

In the meantime the object of this book is to provide a selection of the best aerofoils suitable for models, drawn out in a variety of chords from which the model designer may choose. Geometric data are transmitted, in most publications, as a table of co-ordinates expressed as a percentage of the aerofoil's chord, and a tedious plotting process is necessary before visualisation is possible. This difficulty has been overcome by the use of computer techniques, and 83 sheets of sections have been drawn. In addition to their presentation in 'standard' form, some have been 'skimmed' top and bottom to be suitable as rib outlines, to be covered by sheet or capping strips. Also some sections have been 'blended' to provide outer wing contours in which undersurface camber is gradually removed towards the wing tip.

The Generation of Lift

In describing the nature of air movement as an aerofoil passes through the air, the usual technique is to suppose that the observer moves with the aerofoil. Thus one is faced with describing the flow of air past an apparently stationary aerofoil, as illustrated in *figure 1* and in the photograph on page 9.

Streamlines are defined as imaginary lines drawn so that they lie always in the direction of the local flow: thus it follows that no air crosses a streamline. By this definition the aerofoil itself is a streamline, which is extended upstream and downstream by the so called 'stagnation streamlines' separating the flow which passes above the aerofoil from that which passes below. The aerofoil, set at a small angle of incidence to the flow, curves the streamlines downwards generally, and in its close proximity causes the upper streamlines particularly to bunch together.

In the regions where streamlines come closer together, the air is obliged to flow faster, and thus its kinetic energy increases, proportionally to $(\text{velocity})^2$. This is possible only at the expense of its potential energy (pressure), which drops in these regions. Conversely, where streamlines diverge there is an increase of pressure, and hence a varying pressure field is generated around the aerofoil. The most marked effect is on the upper surface of the nose of the aerofoil where a strong reduction of pressure occurs; on the lower surface there is a general increase of pressure, resulting in an overall force acting upwards and slightly backwards.

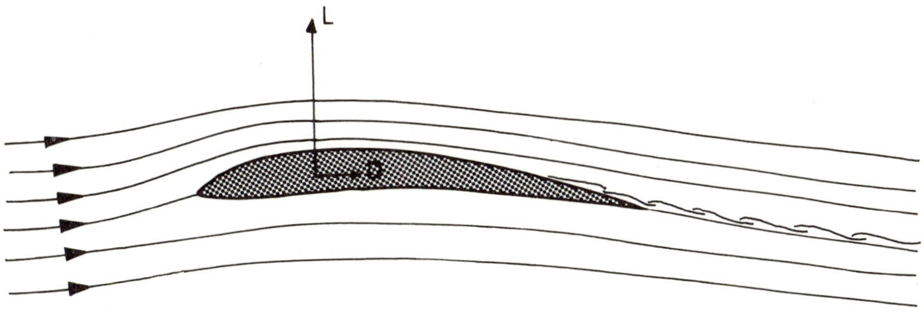

Figure 1 Aerofoil at low incidence, streamline flow.

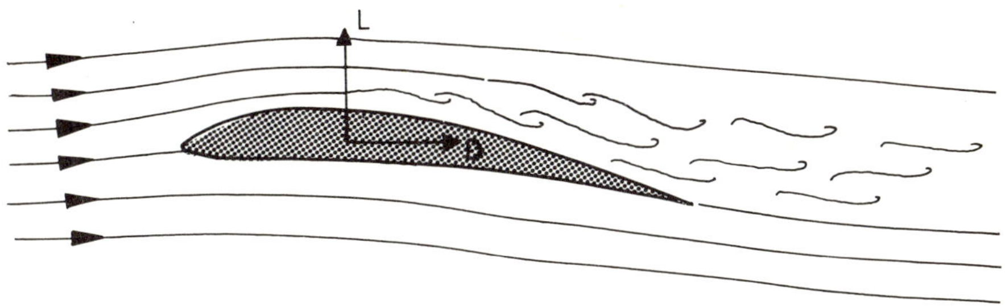

Figure 2 Aerofoil at high incidence, stalled.

The 'lift' force is defined as that component of the overall force acting in a direction normal to the upstream flow; the force component in the tangential direction is referred to as 'drag'. Clearly one is interested in the way the lift and drag of an aerofoil varies with its incidence. At a certain high angle of attack (or incidence), the streamline flow will break down on the upper surface as illustrated in *figure 2*, resulting in a thick turbulent wake accompanied by a drop in lift and a rapid rise of drag. This condition is known as the 'stall'.

At low angles of incidence, forces on the aerofoil increase with incidence and are proportional to (velocity V)^2, the air density ρ, and the aerofoil chord C. It is therefore convenient to non-dimensionalise the lift and drag forces by dividing by these quantities, so that the characteristics of the section are defined by the variation of its lift coefficient C_l and drag coefficient C_d with incidence. Lower case suffixes l and d are used here to denote local aerofoil values, as opposed to overall wing values mentioned later.

$$C_l = \frac{\text{lift per unit span}}{\frac{1}{2}\rho \,.\, V^2 \,.\, C} \qquad C_d = \frac{\text{drag per unit span}}{\frac{1}{2}\rho \,.\, V^2 \,.\, C}$$

The results of some typical wind tunnel tests carried out at Göttingen in 1919 are shown in *figure 3.* The velocity in use and tunnel model construction were not entirely appropriate to model aircraft practice, but the general form of the results is of interest. The lift coefficient increases almost linearly with incidence until the effects of the stall come in, and a maximum lift coefficient ($C_{l\,max}$) is reached. Although drag rises rapidly as the stall is approached, the results include the tip effects of the model test wing and are therefore not entirely due to the aerofoil tested. These wing and tip effects are discussed on page 19.

The results of ingenious tests carried out by S. Suzuki in Japan were published by Frank Zaic in the USA in his *1955–56 Year Book*, and some of these are shown in *figure 4* (page 13). He used a whirling arm apparatus to test wings at model speeds and investigated the effects of construction and roughness.

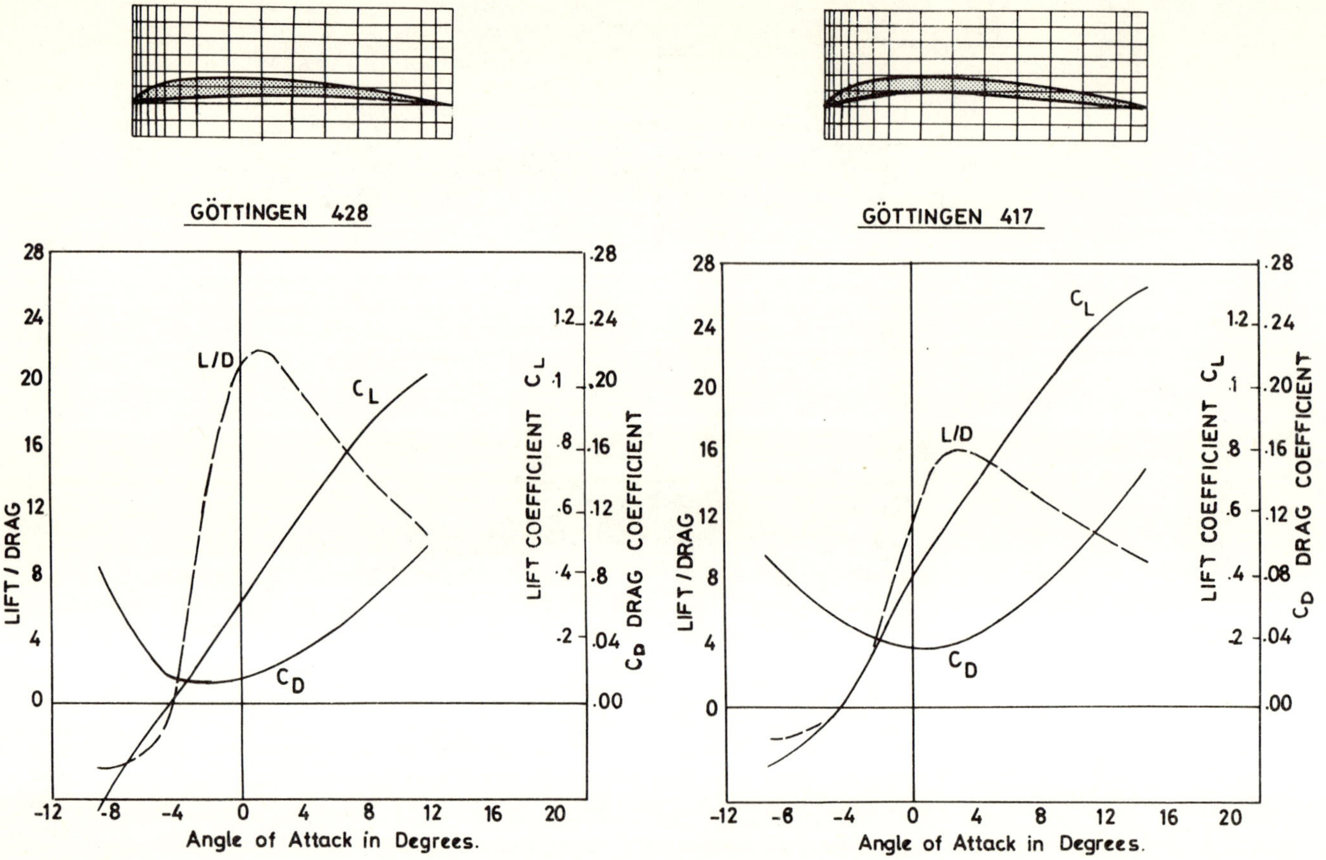

Figure 3 Characteristics of Göttingen aerofoils.

In general one can conclude that increasing camber (arched curvature of the section) improves C_{lmax} and causes the best lift/drag ratio to occur at higher lift coefficients. The effect of upper surface roughness at model speeds is to increase C_l and reduce C_d. The effect of surface spars on the *NACA 6412* is beneficial, increasing both C_{lmax} and lift drag ratio L/D.

Boundary Layers

Air, in common with all fluids, is a viscous substance: that is to say it has a treacle-like quality, but to a very much lesser extent than treacle, of course. Thus air flowing with differing velocities exerts frictional forces within itself, and tangential or shearing forces occur on the surface of the aerofoil. Close to the aerofoil surface, these forces progressively retard the air flow, and the region where this occurs is known as the boundary layer. All drag forces, whether resulting from this surface friction or from the normal pressure distribution, are attributable to these viscous effects. The need to avoid flow separation from any surface of a model and to maintain streamline flow is apparent if one is at all interested in high performance.

The behaviour of boundary layers is complex and not at all well understood under aeromodelling conditions. Initially, at the rounded nose of an aerofoil the boundary layer is laminar, that is to say successive layers flow smoothly over each other. The skin friction is small and the rate of increase of boundary layer thickness is slow. This state of affairs will continue while the local pressure is dropping, but when it begins to rise again the pressure gradient

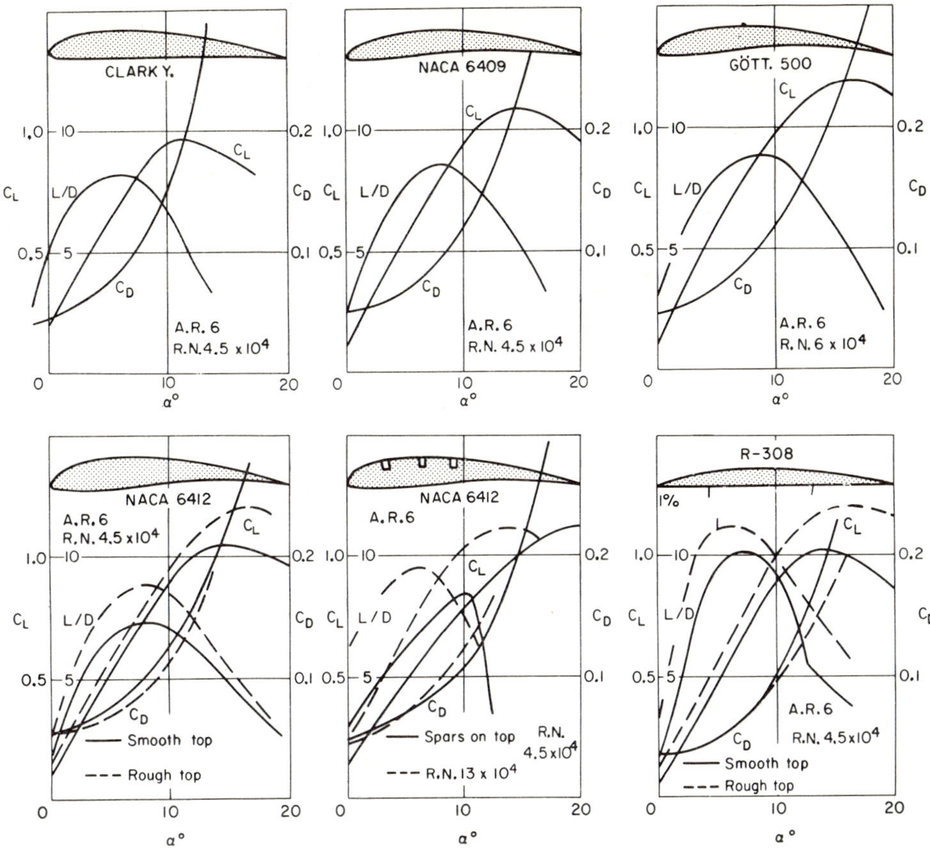

Figure 4 Aerofoil characteristics by S. Suzuki at model Reynold's Numbers. Note effects of roughness and spars.

is termed 'adverse' and the laminar boundary layer will break down and become turbulent. In this region successive layers do not flow smoothly but intermix in a random manner; the skin friction increases, and so does the rate at which the boundary layer thickens. The local air flow in the boundary layer may eventually be retarded to the point where it reverses direction and tries to move along with the aerofoil. The main flow leaves the section at this point, and the space between it and the surface is filled by a highly turbulent wake (see *figure 5*).

The way in which transition from the laminar to turbulent boundary layer occurs is of importance. It is possible for the direction of flow in the laminar boundary layer to reverse prior to becoming turbulent: i.e. a laminar separation occurs. The turbulent boundary layer may then recover enough energy from the mainstream to permit it to reattach itself to the aerofoil, leaving an entrained 'bubble' or vortex on the surface. The bubble may be of varying length, a short bubble occupying a small percentage of the chord and a long bubble extending 30 per cent or more.

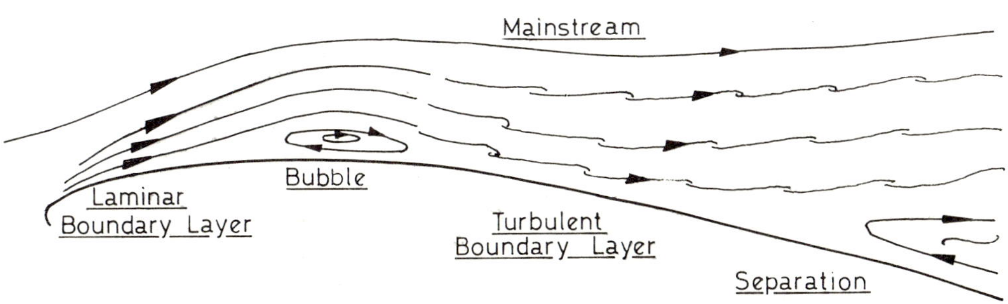

Figure 5 Development of the boundary layer.

A long bubble considerably modifies the pressure distribution around the aerofoil, killing the peak suction at the nose and replacing it with almost constant pressure in the bubble region. When the stall occurs it may do so gradually, as the flow separates progressively from the trailing edge; or it may be sudden due to the laminar flow failing to reattach near the leading edge because the boundary layer bubble has 'burst'.

Reynold's Number

The nature of the flow depends on the relationship between the frictional forces and the inertia forces at work. The boundary layer behaviour depends on this relationship and on the distance it has flowed from the nose of the aerofoil. There is, therefore, an effect of 'scale' and this is measured by a non-dimensional number, known as Reynolds Number, R_e. It depends on the flying speed V, the aerofoil chord C, and the kinematic viscosity of the air. The viscosity is influenced by air temperature and density and is thus dependent on altitude. For modelling purposes these latter effects may be ignored, so that Reynold's Number may be approximately calculated from:

$$R_e = 525\ V \cdot C \qquad V \text{ velocity in ft/sec,} \qquad C \text{ chord in inches.}$$

The effect of Reynold's Number on the lift and drag characteristics of aerofoils at model speeds is very significant. Some MARP results for the *Isaacson 64009* section first published in 1955 are shown in *figure 6.* With reducing Reynold's Number the maximum lift achieved drops and the drag rises. This aerofoil has a fairly sharp nose and its effect is apparent in the curves for R_e = 38,000 and R_e 50,000 above 4° incidence, where the C_l curve steepens and C_d reduces. The leading edge is promoting laminar separation and a bubble forms, giving rise to turbulent reattachment. At R_e = 25,000 the laminar separation does not reattach, and a leading edge stall

Figure 6 Lift coefficient characteristics and lift–drag polar for Isaacson 64009 aerofoil.

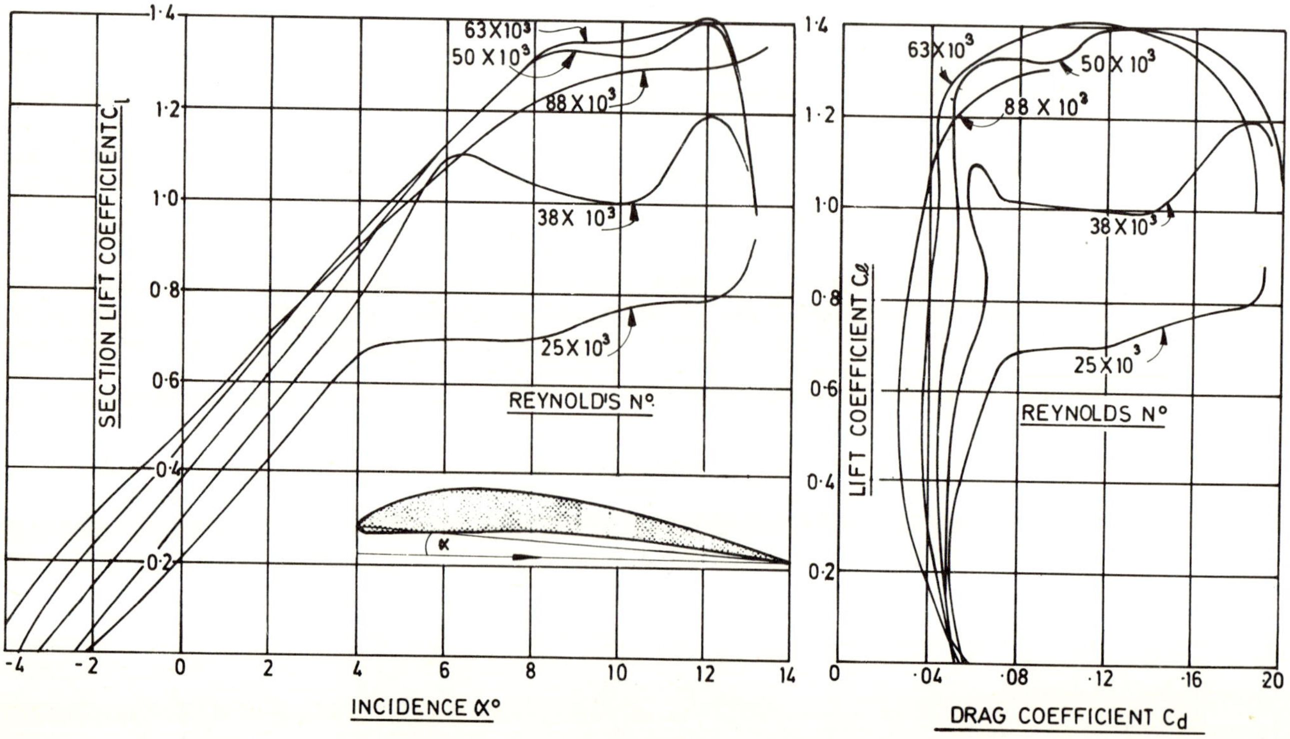

Figure 7 Effect of Reynold's Number on maximum lift coefficient.

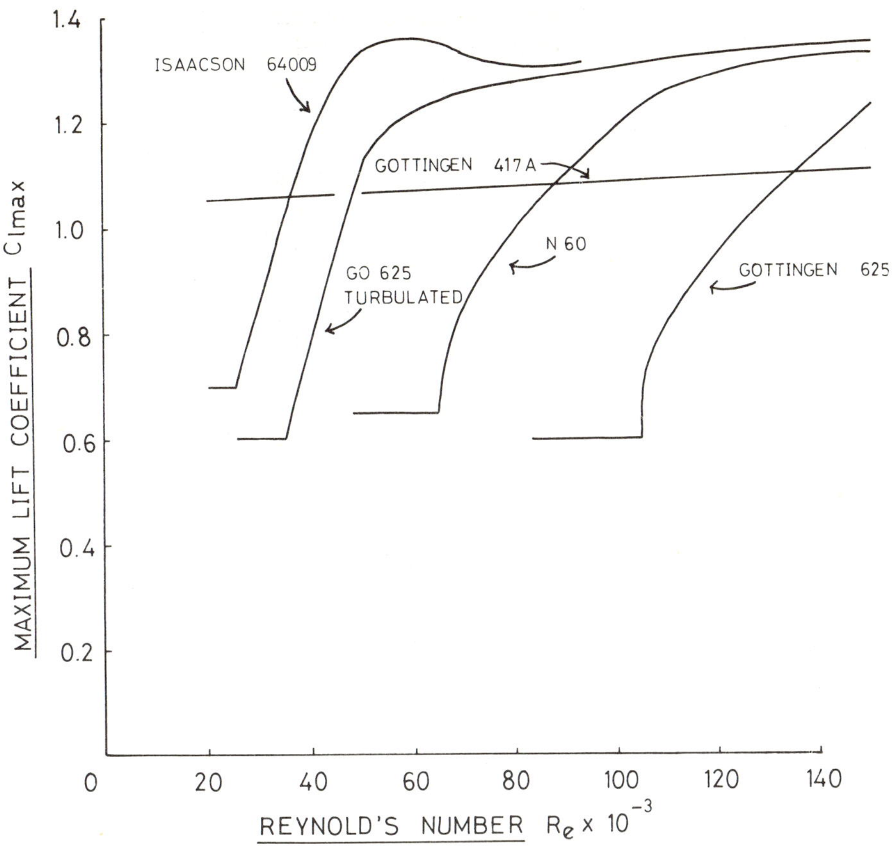

results at low incidence and poor C_{lmax}. This property of the sharp leading edge has been exploited on many successful sections, and its application on Hagel's *F1C* power model is worth noting. This is an unusually highly cambered section for this type of model, which contributes to its gliding performance. At low incidence in the climbing phase the sharp nose probably does not turbulate the boundary layer, and thus minimises drag in this phase of flight.

The result of plotting C_{lmax} versus R_e for the *Isaacson 64009* section is shown in *figure 7* as are other results taken from Schmitz's work. At a certain Reynold's Number known as the critical value C_{lmax} drops rather suddenly: e.g. critical R_e for the *Isaacson 64009* is about 50,000. The *Göttingen 625* and *N60* sections have higher values, but the effect of fitting a turbulator wire in front of the *Göttingen 625* is dramatic, having much the same effect as the *Isaacson 64009* nose. The *Göttingen 417A* is a curved plate of minimal thickness, and although its C_{lmax} is less, there is no critical effect in the range tested. For free-flight models where the attainment of high lift coefficients is so essential, the clear indication is towards the use of thin, cambered sections with turbulent boundary layers.

Lindner's aerofoil, with which he won the world A2 glider championship in 1954, was fitted with a vibrating rubber chord ahead of the leading edge to win again in 1955. Thin strips of wood, or thread doped to the upper surface near the leading edge have also proved successful. Often surface spars are claimed to have the same effect, and recently experiments with double turbulator strips, the second being at about 30 per cent of the chord, show that they seem to help. The second turbulator may be a means of bringing about or stabilising the reattachment of fairly long boundary layer bubbles. Possibly the most effective turbulator known is the use of triangular notches along the upper leading edge. These really constitute a series of vortex generators to break up the boundary layer and promote deep turbulence, and are used by Thomann's and Wortmann's sections, amongst other successful applications. (See photograph on page 34).

Choice of Aerofoil Section

In order to maximise performance, one must select an aerofoil which will provide the lift coefficient required, with due regard to Reynold's Number effects, while minimising drag: the problem is easily stated, but difficult to solve. The table (page 18) shows typical lift coefficients and Reynold's Number's attained by various types of model aircraft. In general, the slow flying free-flight models require high lift coefficients at small Reynold's Numbers, and the faster flying models need low lift coefficients at high Reynold's Numbers. R/C soarers require high and low lift coefficients at intermediate values of R_e. The type of section most suited to each type can best be determined by reference to current practice, as shown in the index to the aerofoils at the end of this book, where the relevant model type is given.

For high speed models the emphasis is on thin streamline sections with smooth surfaces to promote as much laminar flow as possible, and thus minimise drag. For low speed models the emphasis is on thin, cambered, turbulent

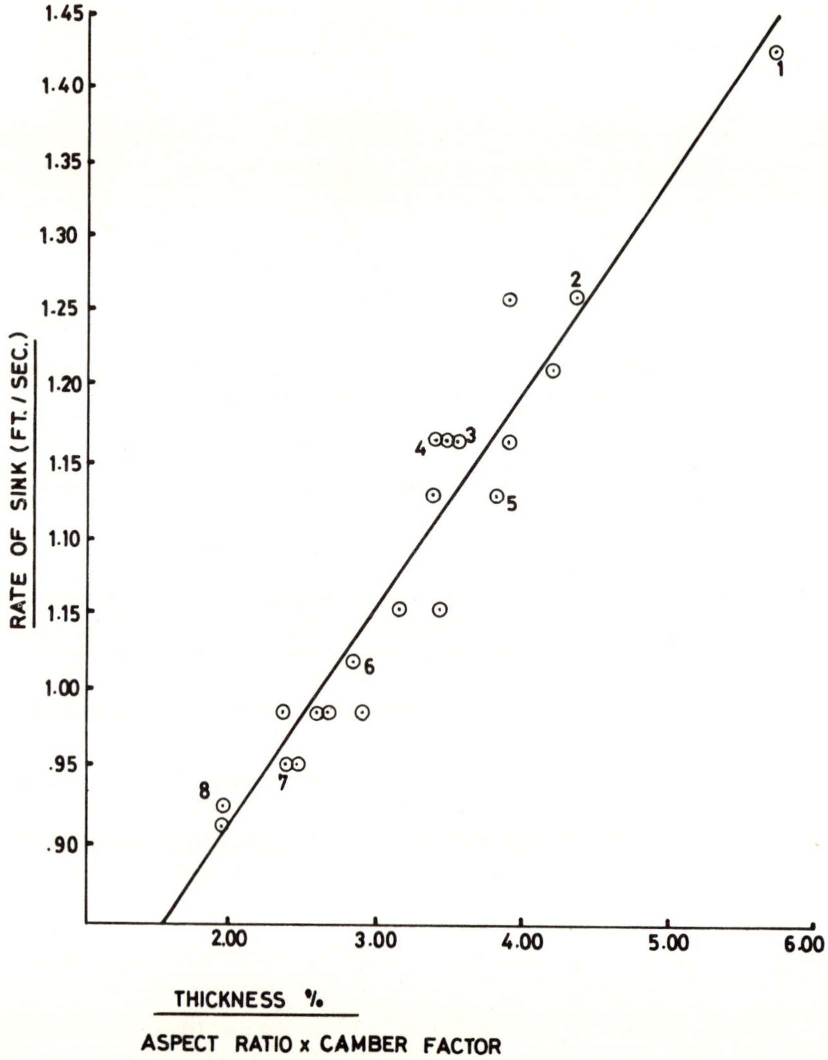

flow sections to promote high lift. For models where wing strength and light weight is critical such as open rubber models and R/C Soarers, a compromise is usually made by using a thicker section of high camber. The lowest Reynold's Number encountered amongst the recognised outdoor classes is 24,000 for Coupe D'Hiver, but its adverse effects on performance are mitigated by the very low wing loadings which can be achieved. The most critical problem arises with the essentially gliding models to International Specifications A1, A2 and Wakefield.

Allnutt and Kaczanowski in their ingenious statistical study, looked at the problem of A2 glider aerofoil selection, and some of their results are given in *figure 8.* They collected data on twenty-one well-proven designs and studied the effect of aerofoil thickness, maximum mean camber, and position of

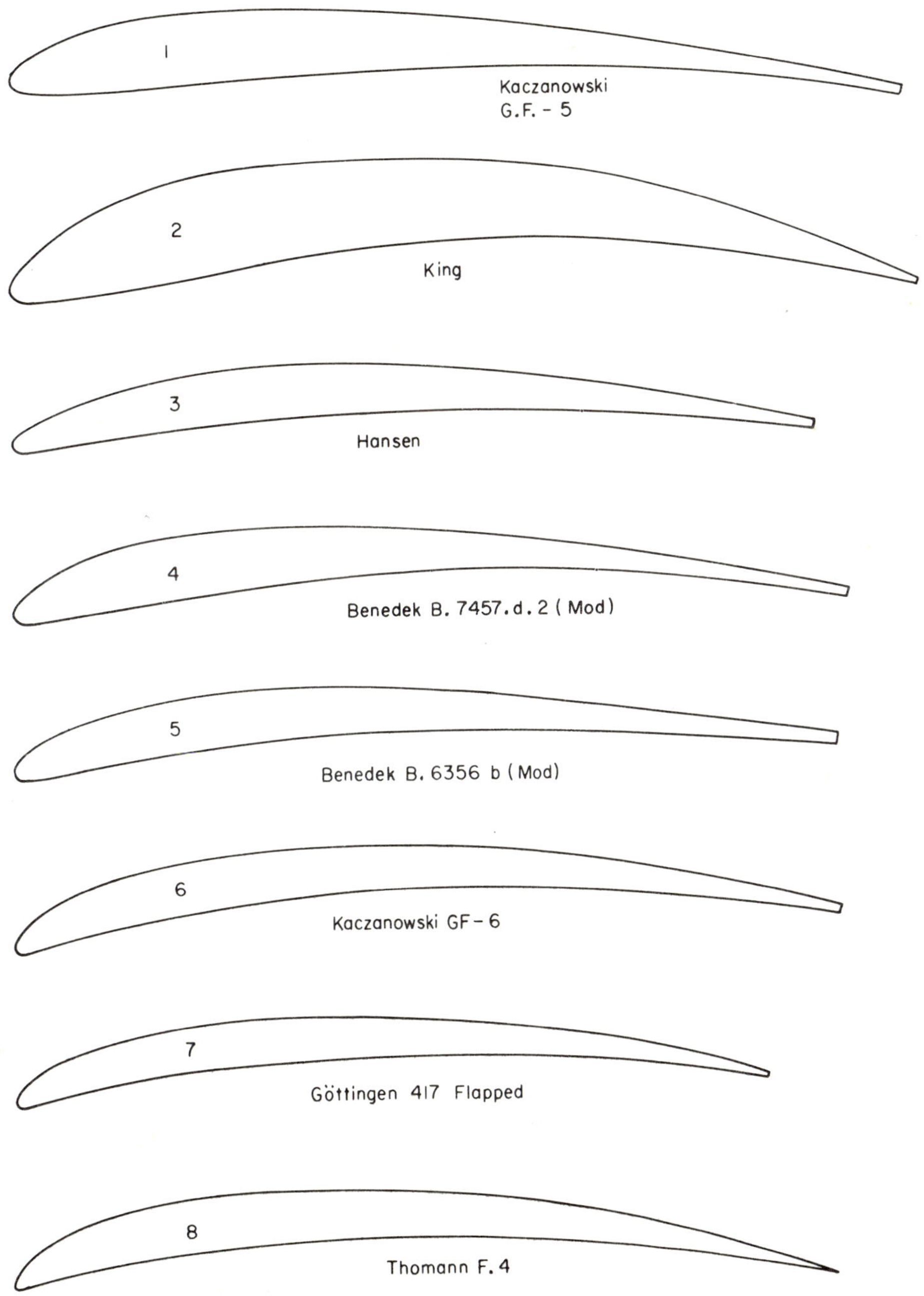

Figure 8 (left and right) Statistical analysis of A2 glider performance by Allnut and Kaczanowski.

Comparison of Typical Lift Coefficients and Reynold's Numbers Attained by Various Model Aircraft in Flight

	Model Type*	Wing Area A in^2	Wing Chord C in	Weight W oz	Wing Loading oz/ft^2	Speed V ft/sec	Lift Coeff. C_L (1)	Reynold's No. $R_e \times 10^{-3}$ (2)
FREE FLIGHT	A2 Glider F1A	445	5·5	14·46	4·67	15·5	1·02	44·7
	Wakefield F1B	247	4·75	8·11	4·73	16·0	0·97	39·9
	FAI Power F1C	442	7·25–4·25(6)	26·45	8·63	50(3)–23·5	0(3)–0·82	190(3)–89·5
	Open Rubber	260	5·5	6·50	3·60	14·0	0·96	40·4
	Coupe D'Hiver	180	4·375	2·26	1·81	10·5	0·86	24·1
	A1 Glider	237	4·75	7·70	4·68	16·0	0·96	39·9
CONTROLLED	Small R/C Soarer	680	7·0	32·0	6·78	18 –70	1·10 –0·07	66·2–257
	Large R/C Soarer	1470	10·0–7·0(6)	80·0	7·30	16(4)–70	1·61(4)–0·08	368–84·0
	C/L Stunt	620	13·5–10·5(6)	44·0	10·22	66	0·12–0·84(5)	467–363(6)
	Rat Racer	105	4·0–3·5(6)	30·0	41·20	191	0·06	400–350(6)
	½A Team Racer	90	5·0–2·5(6)	14·0	22·40	117	0·09	307–154(6)

Notes:
(1) Overall Lift Coefficient $C_L = 7560\ W/V^2 A$
(2) Reynold's Number $R_e = 525\ V \cdot C$
(3) During climb phase with variable incidence tail (VIT)
(4) Level flight using flaps
(5) During 7g manoeuvre using flaps
(6) At the wing tip

*Readers may wish to study the photographs on pages 30–32 to identify the various types of model.

maximum camber, while trying to eliminate the effect of aspect ratio. The aerofoils, drawn in sequence of preference in *figure 8*, clearly indicates the advantage of the thin cambered type, and the influence of extra camber, or flap effect towards the trailing edge is indicated. Unfortunately the extent to which one can extrapolate the result is not known, and we are badly in need of wind tunnel tests to determine the way ahead from here.

Certain types of model have a need to obtain suitable characteristics in more than one operating condition, through a range of lift coefficients and operating speeds. Notable amongst these are free-flight duration power models, R/C soarers and C/L stunt models. While a compromise choice of section can be made, the ultimate performance of these types will almost certainly be obtained through variable geometry sections, involving movable flaps. Slats, plain flaps and slotted flaps have the effect of increasing lift coefficient at the expense of drag while deployed. For C/L stunt purposes high lift coefficients are required in both directions to permit efficient upright and inverted flight. Thus a symmetrical section which changes its camber one way or the other through the use of flaps is standard practice, the simple symmetrical section not permitting high 'g' manoeuvres without the risk of stalling. In the case of R/C soarers the use of flaps can reduce rate of descent in calm conditions or take advantage of weak lift; the flaps could also be used with large deflections as airbrakes in high speed flying tasks, and as lift dumpers in the spot landing manoeuvre. The correct coupling and phasing of the elevator and flap controls will, however, need development work for each application.

For free-flight power, the flap operates in reverse to kill lift and drag by removing camber in the high-speed vertical climb phase, while maintaining the

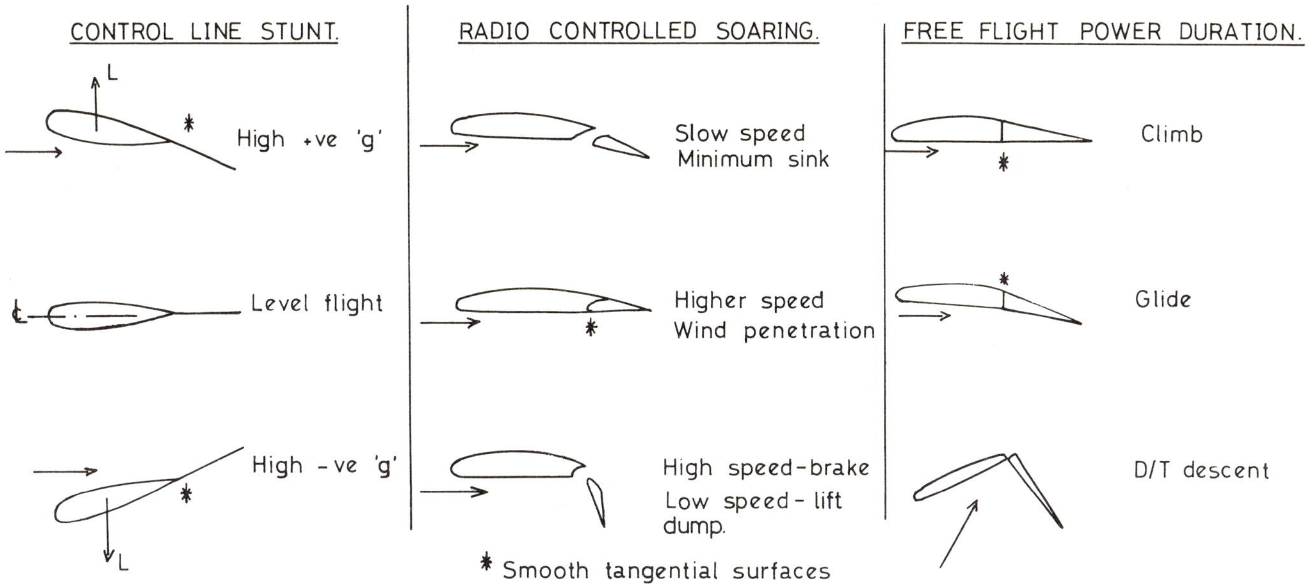

Figure 9 Uses of variable geometry aerofoils.

benefits of camber during the long gliding, slow-speed phase of flight. Experiments have shown that large flaps fully dropped will first stall and then effectively dethermalise a power model or R/C soarer, but this facility has yet to be used by contestants. *Figure 9* illustrates these variable geometry aerofoil uses; see also photographs on page 33.

Wing Effects

The foregoing discussion of aerofoils has treated the matter as viewed in two dimensions, but of course aerofoils are incorporated in wings, and consequently our view of the matter must be extended to embrace three dimensional effects. The passage of a wing through the air, generating lift, causes two vortices to be cast off behind the wing in the direction of flight. This vortex pair can be seen in some circumstances, and is noticeable in the exhaust efflux behind *Concorde*, and aircraft such as *Boeing 707*s which have engines well outboard. Similar swirling effects can be seen by swishing one's hand about in the bath water! The direction of rotation of these vortices is clockwise on the left side and anticlockwise on the right side, as viewed from behind. The vorticity is cast off all along the span, but is very strong towards the tips. The vortex pair which forms a few chords distance behind the wing is often referred to as the 'tip vortices' although their distance apart is a little less than the wing span. These vortices have important effects on wing lift and drag.

In effect the wing vortex system modifies the direction of the flow approaching the wing. Although each section of wing acts substantially in accordance with its two-dimensional characteristics, the effective angle of incidence is reduced over most of the wing. The result is that the wing loading distribution is not generally proportional to wing chord, but degenerates to zero at the wing tips. *Figure 10* illustrates wing spanwise loading distributions for wings of various taper ratios and for rectangular and elliptic planforms. The loading distribution along a rectangular wing is far from constant even for the high aspect ratio value of 16 (i.e., wing span is sixteen times the mean chord). The elliptic wing remarkably produces elliptic loading distribution so that each chord of the wing is equally effective, and this points the way towards wing efficiency.

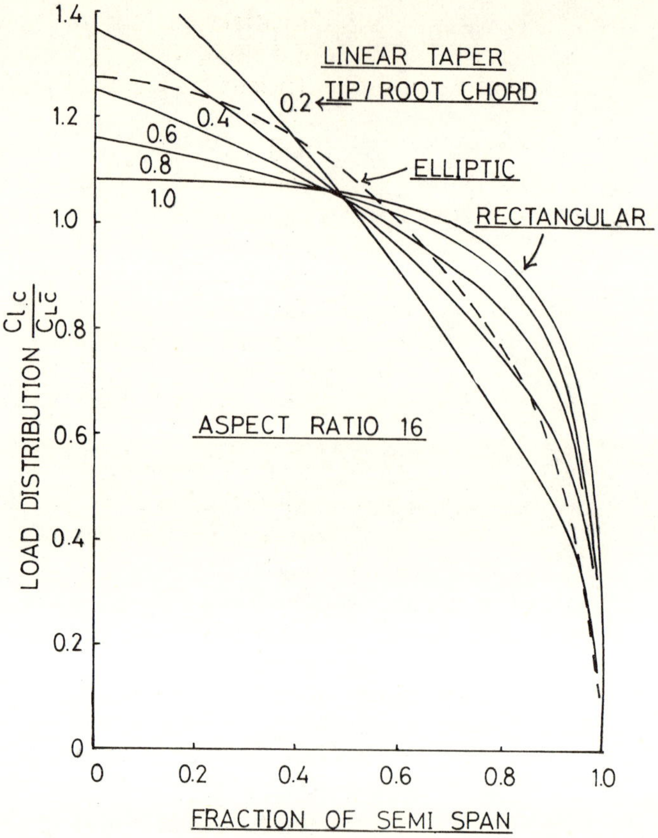

Figure 10 Effect of planform on spanwise load distribution.

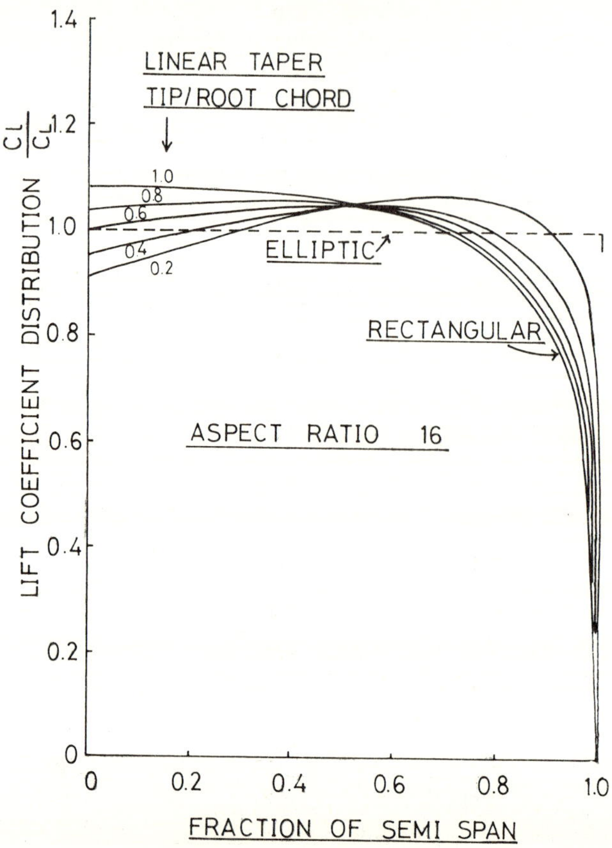

Figure 11 Effect of planform on spanwise lift coefficient distribution.

For wings required to attain high lift coefficients, the limit arises when the stall sets in at some point along the wing. *Figure 11* illustrates wing spanwise distribution of lift coefficient for wings of various taper ratios, and for rectangular and elliptic planforms. For rectangular wings $C_{l max}$ occurs at the wing root. A straight stall usually results without a strong tendency for a wing to drop, and outboard ailerons (if they are in use) remain effective. For wings of high linear taper, $C_{l max}$ occurs at about 70 per cent of the semi-span. In the event of a stall there is a strong wing dropping tendency because one side will inevitably stall before the other; and ailerons, if deployed normally to correct this, would in fact deepen the stall and worsen the effect. The elliptic wing has a constant lift coefficient distribution and can therefore attain the highest overall lift coefficient ($C_{l max}$), but its stall would be simultaneously complete. See *figure 12.*

The wing vortex system requires energy to sustain it, and this results in an additional drag force on the model aircraft. It comes about as a result of the local wing incidence being reduced by the vortex downwash. The lift force is inclined backwards through this angle and therefore acquires a component in the dragwise direction. This lift dependant drag is known as 'induced drag'. Its value depends on the (lift coefficient)^2, the wing planform shape, and is inversely proportional to aspect ratio. In non-dimensional terms, the induced drag coefficient can be expressed:

$$C_{Di} = \frac{k \cdot C_L{}^2}{\pi \text{ aspect ratio}}$$

The induced drag factor k is dependent on planform and its variation may be studied by reference to *figure 13.* Its minimum value is one, which is achieved only by a wing with elliptical loading distribution, such as an elliptic planform itself, or another planform with appropriate twist. A rectangular wing is worst in respect of induced drag, and if linear taper from root to tip is to be used, then a taper ratio of about 4 : 1 is best. Clearly induced drag is of the greatest importance for models operating at high overall lift coefficients. For example the induced drag of an A2 glider is 40–50 per cent of its total drag, and the choice of wing planform and aspect ratio is very important.

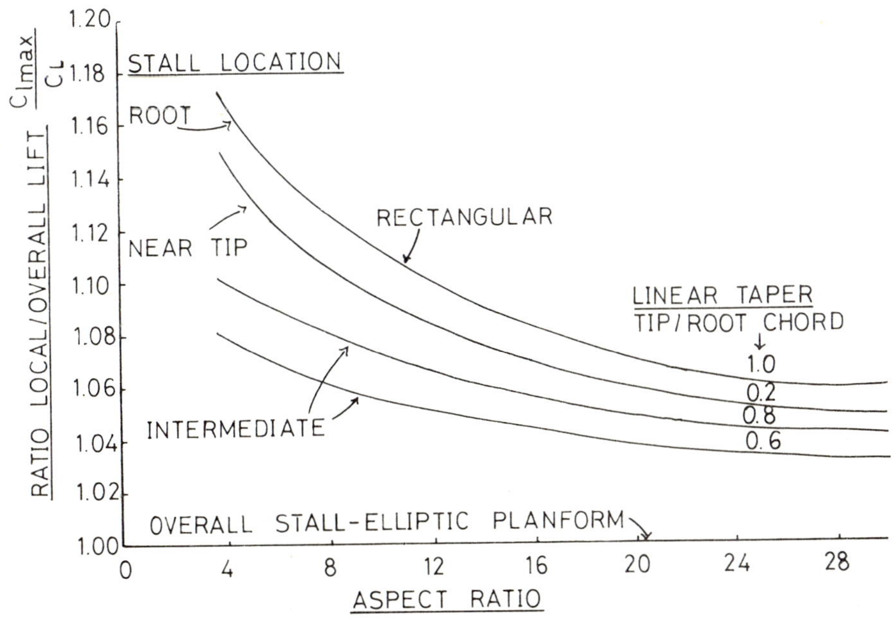

Figure 12 Effect of aspect ratio on lift ratio for wings of various planforms.

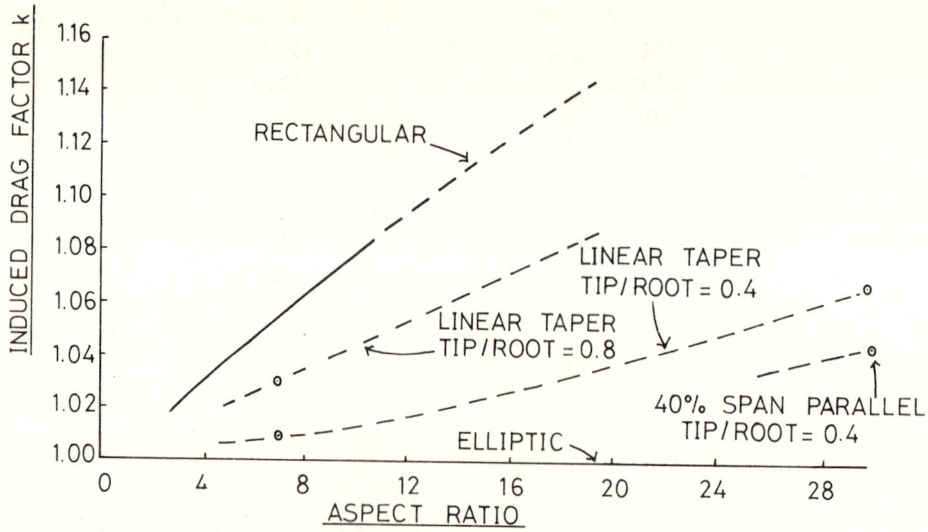

Figure 13 Effect of aspect ratio on induced drag factor for wings of various planforms.

For slow-flying gliding models, the best choice of planform for performance is not quite clear. On the one hand one may choose a shape as close to elliptic as construction methods permit. A wing in which the central 40 per cent is parallel, and the remainder tapers to a tip chord 40 per cent of the root chord, is an excellent compromise and may be observed on several full-size gliders, man-powered aircraft, and other types designed to operate at higher lift co-efficients. This would seem the best choice for R/C soarers and large models, providing the Reynold's Number towards the tip does not approach its critical value. On the other hand, for smaller models, such as the Wakefield, rather less taper seems appropriate to keep Reynold's Number above critical. The A2 glider is in a state of gradual change. In Great Britain, the rectangular wing with rounded tips has been the norm for many years. However, under mid-European influence more tapered wings are in evidence. A2 glider performance is looked at further on the next page.

Wing tip shape is probably quite important, but it is here that the greatest variation is observed, together with fin outline shape. Wing tip shape is one of the most aesthetically distinguishing features of a model and there is a reluctance amongst designers to standardise on these features. Generally wing loading must drop to zero at the tip, although vortex upwash can produce areas of peak suction on some wings. There is a tendency for air to flow around the tip from the region of high pressure beneath to low pressure above. Because the tip produces little lift, it must be at low aerodynamic incidence, and a section with small or zero camber would seem appropriate. Certainly an undercambered lower surface should be avoided at the wing tip to prevent boundary layer separation on the undersurface. Some designers favour blending out undercamber for this reason and because of the favourable effect it has on loading distribution. The wing tip of the Britten Norman *Islander* aircraft was specially developed to suit the rectangular wing and is worthy of study by modellers. There is evidence that the wing span should reach its maximum at the trailing edge, and raked tips on models are quite common. The cross-section of tips is normally rounded, although on the Short *Skyvan* fin and Handley Page *Jetstream* tail the span was extended two or three inches by fitting flat triangular plates known as 'vortex concentrators'. These certainly improved control effectiveness. Conical camber tips, which can be seen on several of the latest Cessna aircraft, including the *Aerobat, Skyhawk* and

Cardinal, and on the Fairchild *A-10A,* have been tried on models with apparent benefits. Vertical tip plates are not recommended because these in effect have two tips of their own and produce excessive drag when yawed (see photograph on page 34).

Gliding Performance

It will have become clear from the discussion above that the prediction of gliding performance is indeed rather difficult to achieve with accuracy. Designing a model for optimum gliding performance is a complexity which we should like to unravel. In full-size aviation accurate results are only achieved after the extensive wind tunnel testing of individual projects, backed up by flight test experience. However, let us look at what can be done.

The average rate of descent 'u', of a gliding model can be expressed as:

$$u = 87 \left(\frac{W}{A} \cdot \frac{C_D{}^2}{C_L{}^3}\right)^{\frac{1}{2}} \text{ ft/sec where } \frac{W}{A} \text{ is the wing loading in oz/in}^2$$

This equation is deceptively simple in appearance, but does clearly embrace the two decisive parameters of the problem: the physical characteristics of the model giving rise to its wing loading, and the aerodynamic characteristics giving rise to lift and drag. We are concerned here with the aerodynamic characteristics; we do not wish to minimise the importance of the physical difficulties of model design, but these could be the subject of another publication.

Usually the aeromodeller faces the problem of minimising $C_D{}^2/C_L{}^3$ when he comes to trim his model. The best trim is found close to the stall; the technique is usually to obtain a slightly stalling flight and then back off a little to obtain smooth flight but with a slight 'liveliness' still noticeable. Free-flight models are trimmed to turn, and even wings with central stalling tendencies will drop the inner wing during a stalled turn, presumably due to the lower Reynold's Number appertaining to that side. For some aerofoils it is apparent that the trim can be too close to the stall, resulting in flight in the high drag region without actually losing lift. The resulting type of flight is referred to as 'mushing'. It seems that models exhibiting this are sometimes capable of remarkably good performances in calm conditions, but mush when flown in windy, turbulent conditions.

A2 Glider Example Calculation

The performance of an A2 glider with a rectangular wing is to be estimated. We will assume the characteristics of the *Isaacson 64009* section, but recognising that modern aerofoils for gliders are somewhat thinner we will reduce the drag values to 80 per cent. Further, we will work to the highest local lift coefficient attainable relative to the Reynold's Number, but with a reduction to 95 per cent in recognition of trimming technique. The A2 glider will have a wing area $A = 445$ in^2, and is required to weigh $W = 14{\cdot}46$ oz in total.

From figure 7 with $R_e = 50{,}000$
$C_{l\,max} = 1{\cdot}34$ therefore $C_l = 0{\cdot}95 \times 1{\cdot}34 = 1{\cdot}27$

From figure 6 with $C_l = 1{\cdot}27$
$C_d = 0{\cdot}05$ therefore $C_d = 0{\cdot}80 \times 0{\cdot}05 = 0{\cdot}040$

Now $R_e = 525\ V.C$ therefore $V = \dfrac{50000}{525C} = \dfrac{95{\cdot}3}{C}$

And overall lift coefficient

$$C_L = \frac{7560 \cdot W}{V^2 \cdot A} = \frac{7560 \times 14{\cdot}46}{95{\cdot}3 \times 95{\cdot}3} \times \frac{C^2}{A} = 12\frac{C^2}{A}$$

But the aspect ratio $a = A/C^2$ therefore $C_L = \dfrac{12}{a}$

The aspect ratio is found by a step by step calculation:

Guess aspect ratio a	10·0	10·5
Find C_l/C_L from fig. 13	1·160	1·108
Find C_L for C_l = 1·27	1·095	1·148
Find aspect ratio a = $12/C_L$	10·95	10·45

Thus to a close approximation aspect ratio a = 10.5

From *figure 13* find induced drag factor k = 1·082

$$C_{Di} = \frac{kC_L^2}{\pi a} = \frac{1{\cdot}082 \times 1{\cdot}148 \times 1{\cdot}148}{\pi \times 10{\cdot}5} = 0{\cdot}043$$

Total drag, including allowance for fuselage and tail 0·005 say

$$C_D = 0{\cdot}040 + 0{\cdot}043 + 0{\cdot}005 = 0{\cdot}088$$

Rate of descent

$$u = 87\left(\frac{W}{A} \cdot \frac{C_D{}^2}{C_L{}^3}\right)^{\frac{1}{2}} = 87\left(\frac{14{\cdot}47}{445} \times \frac{0{\cdot}088^2}{1{\cdot}148^3}\right)^{\frac{1}{2}} = 1{\cdot}123 \text{ ft/sec}$$

And finally the average duration from 164 ft (50 metres) towline

$$= \frac{164}{1{\cdot}123 \times 60} = 2 \text{ min } 26 \text{ sec}$$

The aspect ratio is 10·5, which means the wing span is 68·3 in projected, and the chord is 6·52 in.

The result of this calculation is quite encouraging and what strikes one on looking back at figure 7 is that a lower initial lift coefficient could be tried, in exchange for lower critical R_e and increased aspect ratio. This has been done and the result is given in *figure 14.* The result emerges for rectangular wings that the optimum aspect ratio is about 14 : 1, producing a duration of 2 mins 32 sec. This result is very much in line with current practice, but the fairly flat nature of the optimum is demonstrated.

After trying the calculation in this way, wings of other planforms were studied. The apparent advantages of the elliptical wing have been mentioned and the practical form of linear taper to 40 per cent of chord over the outer 30 per cent of span was studied first. The critical region for stalling is at about 70 per cent of semi-span where the chord is 70 per cent of the root value. This small chord limits the aspect ratio and detracts from the beneficial effects, resulting in a performance that is significantly worse than for rectangular wings.

Having reached this stage a criteria for optimum wing planform emerges. The taper should be such that the resulting C_l distribution matches the $C_{l max}$ locally available with regard to the spanwise distribution of Reynold's Number.

Figure 14 The predicted effects of planform and aspect ratio on A2 glider performance.

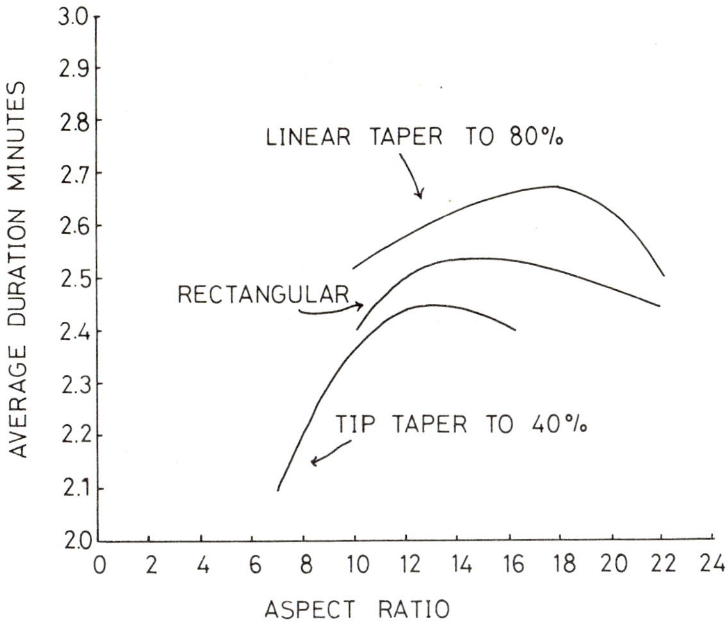

A good approximation to this is found using a linear taper to 80 per cent of root chord at the wing tip. This wing stalls first at about 40 per cent semispan where the chord is 92 per cent root chord, so that the benefits of aspect ratio are not unduly spoilt, while making the most of the other advantages. The results given in *figure 14* shows a duration of 2 mins 40 secs attained with an aspect ratio of 18. The best known glider of this type is the Hafsass/Spooner *SPL* design, for which a duration of 2 mins 57 secs has been claimed.

Thermal Soaring Glider Example Calculation

A large thermal soaring glider is to be designed with a wing area of 1300 in^2 and aspect ratio of 16 : 1, for which the weight is estimated to be 70 oz. The wing section chosen is the *Eppler 385*, but the *Isaacson 64009* for which data is provided in Fig. 6 is not untypical. A comparison of wing planform effects is called for.

The model is large enough to be clear of critical Reynold's Number effects and one can expect to obtain an overall lift coefficient $C_L = 1{\cdot}2$ in conjunction with generously tapered wings.

$$C_L = \frac{7560\ W}{V^2\ .\ A} \quad \text{therefore} \quad V = \left(\frac{7560 \times 70}{1{\cdot}2 \times 1300}\right)^{\frac{1}{2}} = 18{\cdot}4 \text{ ft/sec}$$

Average Reynold's Number with mean chord $C = (1300/16)^{\frac{1}{2}} = 9$ in
$Re = 525\ V\ .\ C = 525 \times 18{\cdot}4 \times 9{\cdot}0 = 87{,}000$
Figure 6 gives aerofoil characteristics at $R_e = 88{,}000$ which is sufficiently typical for this calculation.

Planform 1:

Linear taper, tip chord = 40 per cent root chord

From Fig. 12, $C_{l max}/CL = 1{\cdot}05$ and for $C_l = 1{\cdot}2$, $C_L = 1{\cdot}14$
From Fig. 14 Induced drag factor $k = 1{\cdot}03$

$$\text{therefore } C_{D_i} = \frac{kC_L{}^2}{\pi a} = \frac{1{\cdot}03 \times 1{\cdot}14^2}{\pi \times 16} = 0{\cdot}027$$

From figure 6 at $C_l = 1{\cdot}2$, $C_d = 0{\cdot}050$ and allowance of $C_d = 0{\cdot}010$ will be included for fuselage and tail

$$C_D = 0{\cdot}050 + 0{\cdot}027 + 0{\cdot}010 = 0{\cdot}087$$

$$\text{Rate of descent } u = 87\left(\frac{W}{A}\cdot\frac{C_D{}^2}{C_L{}^3}\right)^{\frac{1}{2}} = 87\left(\frac{70}{1300} \times \frac{0.087^2}{1{\cdot}14^3}\right)^{\frac{1}{2}} = 1{\cdot}44 \text{ ft/sec}$$

Duration off 492 ft (150 m) towline $= \dfrac{492}{1{\cdot}44 \times 60}$ = **5 min 41 sec**

Planform 2:

40 per cent span parallel, tip chord = 40 per cent root chord.

This is a practical approach to the elliptical wing and one may take $C_{l\,max}/C_L = 1{\cdot}0$ and $k = 1{\cdot}0$

$$\text{Induced drag } C_{Di} = \frac{k\,C_L{}^2}{\pi a} = \frac{1{\cdot}0 \times 1{\cdot}2^2}{\pi \times 16} = 0{\cdot}029$$

$$\text{Total drag } C_D = 0{\cdot}050 + 0{\cdot}029 + 0{\cdot}010 = 0{\cdot}089$$

$$\text{Rate of descent } u = 87\left(\frac{W}{A}\cdot\frac{C_D{}^2}{C_L{}^3}\right)^{\frac{1}{2}} = 87\left(\frac{70}{1300} \times \frac{0.089^2}{1{\cdot}20^3}\right)^{\frac{1}{2}} = 1{\cdot}36 \text{ ft/sec}$$

Duration off 492 ft tow $= \dfrac{492}{1{\cdot}36 \times 60}$ = **6 min 2 sec**

The linear taper chosen as *planform 1* is about the best choice amongst the linear tapers, and is not greatly deficient in comparison with *planform 2* in terms of average duration. However, the difference in performance might be critical in a contest in calm conditions, when attempting to score the maximum duration set at 6 mins under class F3B rules for the duration task. In windy conditions, of course, the flying speed must be increased in order to make headway, with consequent reduction of the lift coefficient. In these conditions the induced drag becomes a smaller proportion of the total drag and wing planform effects are less important.

From the point of view of aerofoil choice and gliding performance, duration and distance tasks make very similar demands. Distance may be estimated from the equation:

$$\textbf{Distance} = \textbf{Launch height} \times \frac{C_L}{C_D} = 150 \times \frac{1{\cdot}20}{0{\cdot}089} = 202 \textbf{ metres}$$

In practice some of this distance must be used for turns at each end of the course, and the remainder will only be realised under calm conditions. However *planform 2* has a potential advantage over *planform 1* of about 50 metres.

Use of Aerofoil Sheets

The eighty-three pages of aerofoil sections which follow have been selected as those best representing current trends in the various applications for free-flight, radio control and control line model aircraft. Each section has been drawn in a range of chord sizes appropriate to its application and measured in millimetres. The metric size for the sections is used because wood sizes now come in mm units and the Federation Aeronautique Internationale (FAI) model specifications are similarly phrased.

To aid the use of the data, each section has its chord line and its 30 per cent datum drawn, and these are useful when laying out tentative structural

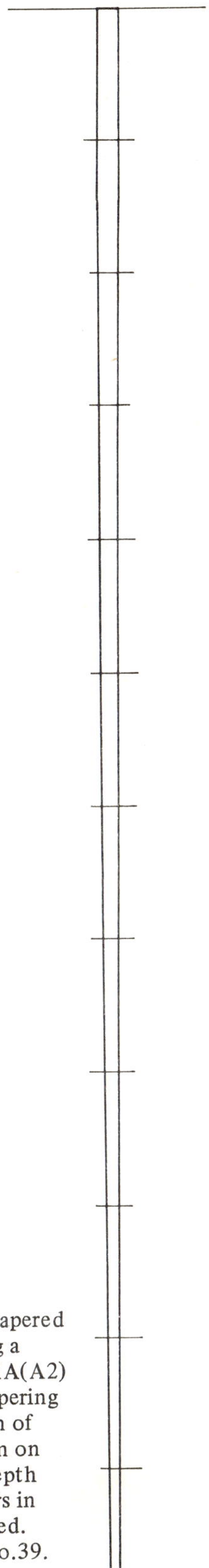

Figure 15 Example of tapered wing tip structure using a blended aerofoil—an F1A(A2) glider. Note the spar tapering and the discontinuation of one flange. The diagram on the right permits the depth of the slots for the spars in the ribs to be determined. See Aerofoil Section No.39.

arrangements. In addition each page has an inch scale drawn on it with which one can check for paper shrinkages or expansion.

Some sections are drawn 'skimmed'. That is, 1·5 mm or 0·8 mm has been removed from the outer surface to give accurate rib shape for cases where cap strips or full sheeting of the profile is to be employed. Other sheets have been prepared with 'blending'. That is, a family of sections have been drawn gradually removing camber to suit outer wing contours.

By first referring to the index of aerofoil sections at the end of the book, the model designer may quickly visualise the range available here for his use. By tracing or drawing directly on the pages, tentative structural arrangements may be tried out until a satisfying design is evolved. Some examples of model structures are included in *figures 15* and *16.*

Finally, templates of the desired sections can be obtained from tracings or by pricking through the pages of the book. Cutting out individual sections or pages is not recommended because this will destroy the permanent record of sections at your disposal.

There is a descriptive index to the aerofoil sections at the end of this book

Figure 16 Examples of model wing structures using standard or skimmed aerofoils.

(a) A1 glider structure based on *Benedek 6407E* section with sheeted surfaces and exposed sub-ribs.

(b) F1A (A2) glider structure with tapering spruce spars, full chord cap strips on the ribs, and intercostal spar webs and turbulator strips, using *Lindner* section.

(c) F1C power structure with top and bottom sheeted surfaces, and tapering spruce insert in top surface, using *Göttingen 795.*

(d) Coupe D'Hiver structure using *Koster 66* section. Close pitched balsa spars promote turbulence, with wide pitched ribs.

(e) F1B (Wakefield) structure using *Swartzbach 68* section. I-section spar and sheeted top nose surface. A thread turbulator is fitted.

(f) Open rubber model structure using *Isaacson 64009* section. I-section main spar with vertical web between ribs. Secondary spar for turbulence.

(g) R/C soarer structure for large model using *Eppler 387* section. Twin I spars in balsa, reinforced by spruce in centre section. Fully sheeted nose box, with capped ribs.

(h) C/L stunt structure based on *NACA 0016* section with extension for flap. I-section spar with sheeted nose box and capped ribs with weight saving cut-outs.

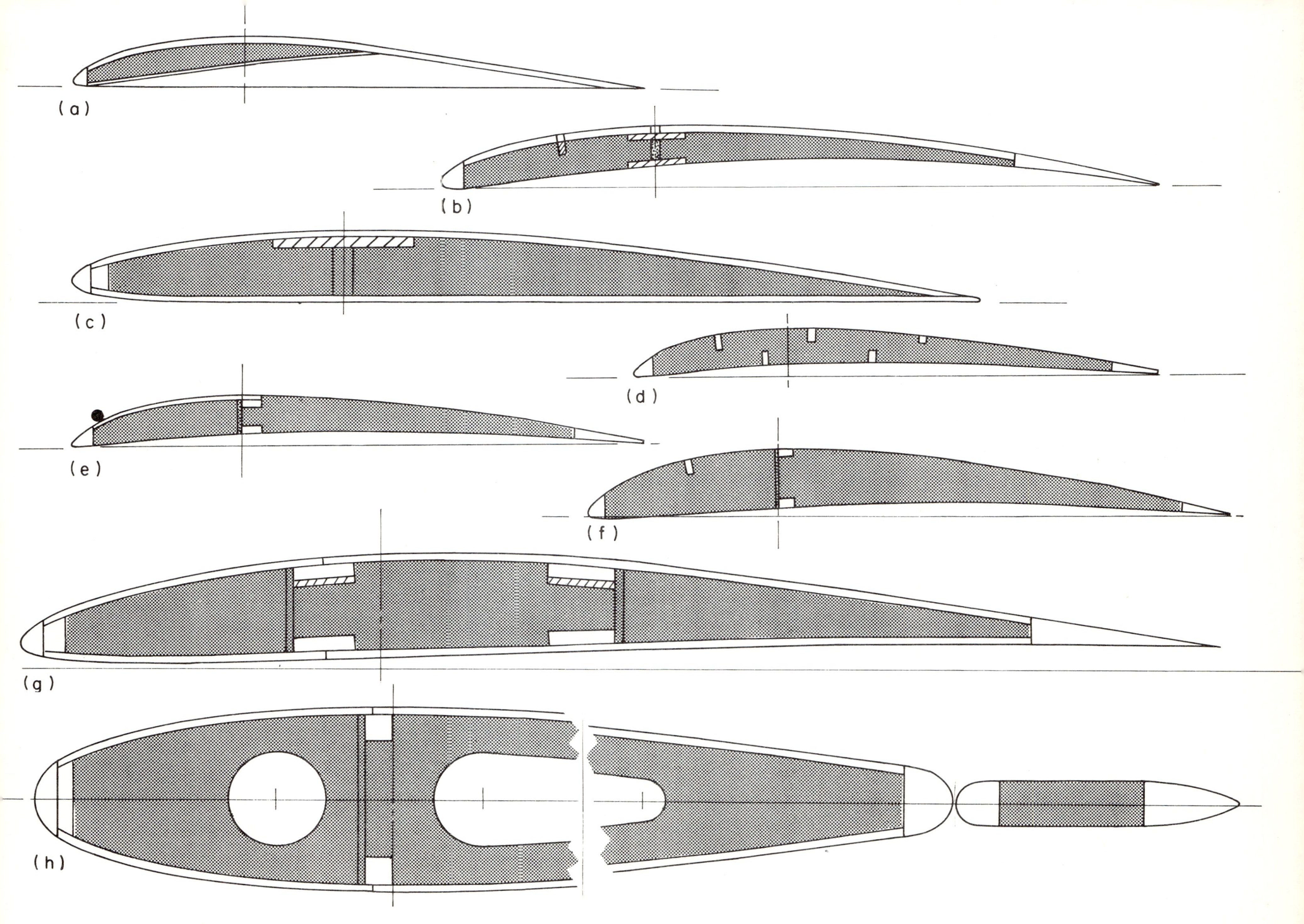
(a)
(b)
(c)
(d)
(e)
(f)
(g)
(h)

FREE-FLIGHT MODEL AIRCRAFT
(see also: Frontispiece)

Right Hans Seelig is chaired to victory, displaying his (F1C) power duration model. His name is synonymous world-wide with the clockwork timers he produces to operate model mechanisms.

Above Rather attractive models of Jon Davis of the USA. In the foreground is the International class (F1A) sailplane and being held, the (F1B) Wakefield rubber driven model.

Left Coupe D'Hiver, the small class rubber driven model, displayed by Guilio Gastaldo, 1973 Champion of Italy.

Right George French representing Great Britain at an international meeting, about to launch his F1C power model. George was one of the pioneers, in Great Britain, of the variable incidence tailplane.

Above left Speed model by Dave Smith of Great Britain. A detachable wing tip covers the line attachments to reduce drag.

Above right A British 'Goodyear' racer, with exposed engine, tank and controls, as permitted by the rules to provide a simplified form of team racing.

CONTROL-LINE MODEL AIRCRAFT

Right A highly developed Italian speed model. The wing is placed on the inside of the fuselage where it flies at a slightly slower speed than if it were symmetrically positioned. The tailplane is positioned on the outside where it has clean air in which to operate.

Below An attractive and unusual control line stunt model by P. Rocher of France. The tip fins could be expected to improve elevator effectiveness.

Below right The team race rules call for a semi-scale type of model. Alan Hill's model is clearly based on the classic lines of the Spitfire. The elliptical shape of the wings and tailplane help to minimise the model's drag.

Right A beautiful scale Spitfire. Models such as this, equipped with retracting undercarriage, throttle and all main flying controls, can perform all the manoeuvres of the original full-sized machines.

Left A collection of immense, thermal soaring, model gliders, as used in competitions involving duration, distance, speed and spot landing tasks.

Terry Cooper, British Team Member, displays his stunt model. Such models must perform as well inverted as right way up, and utilise symmetrical wing sections.

A pylon racer, capable of high speed in straight and turning flight. The sharp leading edges of the wing and tail surfaces can be seen.

Above Thomas Koster of Denmark prepares to launch his F1C power duration model 'Grootna'. The wing is equipped with a flap to achieve a thin uncambered wing section for the high speed climb.

For the glide phase of flight, the flap lowers a few degrees to improve lift and stretch the glide duration. The tail-plane incidence is also varied to maintain the trim and eventually 'pops up' to dethermalise and terminate the flight, in which position it is seen.

Left Wing flaps operated by radio control. In this case the hinge position is below the wing, so that the flaps effectively increase wing camber and wing area on being deployed. The same principle is found on large, full-size aircraft as a means of minimising flying speed for take-off and landing.

VARIABLE GEOMETRY FEATURES

Right Gary Phelps' (USA) control line stunt model demonstrates the differential movement of its wing flaps and elevators. The deflections shown would suit an outside loop manoeuvre (bunt), where the tail is lifted by the elevator and wing lift (down-wards) is augmented by the flaps.

AEROFOILS IN ACTION

Right The A2 glider of Tahkapaa from Finland, demonstrating its pylon mounted wing. This is used to minimise the interference of the fuselage with the airflow past the wing.

Far right The upright engine installation of Urs Schaller's Swiss F1C power model is faired in by a nacelle which merges with the wing supporting pylon.

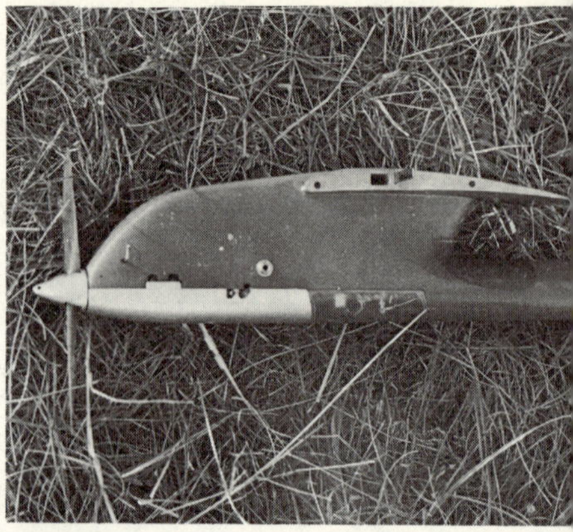

Left The wing attachment features of this pair of A2 gliders illustrates the slim, cambered shapes of modern sailplane aerofoils.

Right Jedelsky type of wing construction, using a solid balsa nose, with thin sheet cambered aft section, supported by exposed ribs on the underside. The Benedek 6407E was developed in conjunction with this method of construction.

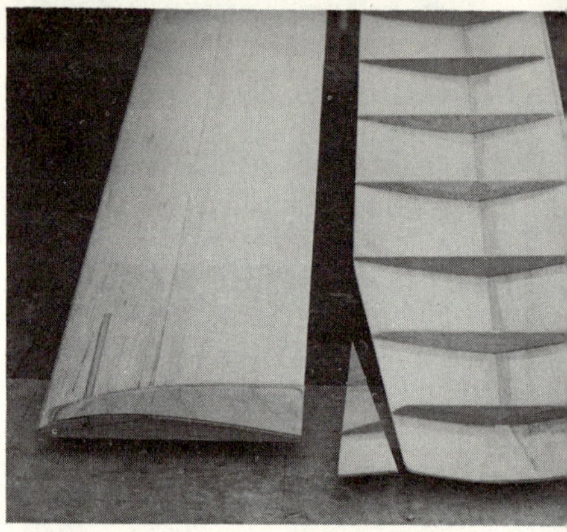

Right Thomann's model illustrates the Flamingo wing section. The advantage of providing depth to accommodate the wing spars, and for the box and tongue attachment is apparent.

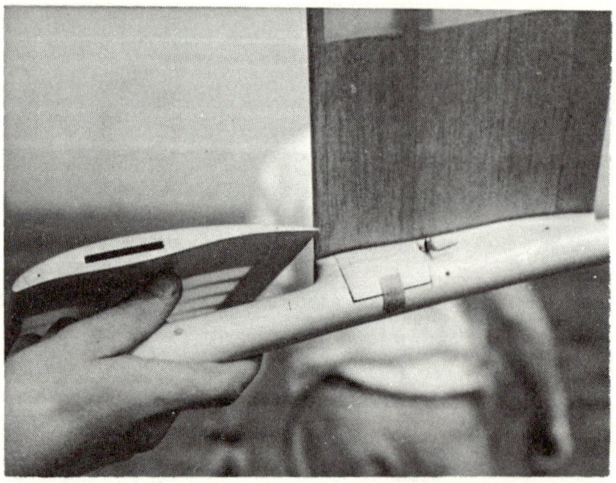

Above Triangular notch turbulators as seen on a pair of A2 glider wings (see page 16). *Photograph: Bryan Spooner.*

Left Britten Norman Islander aircraft. Wing tip configuration has upswept trailing edge producing tip washout.

Right Piper Cherokee Six aircraft, fitted with conically cambered wing tips.

Above left Gerry Ritz of the USA with his high aspect ratio A2 glider design with which he won the world championship. His wing section is very slim and wing structural strength depends entirely on the leading edge and wide trailing edge, no spars are utilised.

Above Rolf Hagel of Sweden with his F1C power duration model. His use of sharp nosed, undercambered sections is a little unusual for this class of model, but their performance is undisputed.

Left George Benedek of Hungary, whose work on aerofoils for model aircraft is renowned throughout the world.

MODEL AEROFOIL DESIGNERS

Below left Oscar Czepa of Austria, another world champion, the first to win using a relatively modern, slim section with his very long 'toothpick' design.

Below Max Hacklinger, with his A2 glider which features rubber cord turbulators ahead of the leading edge. In flight these buzz and produce turbulence in the flow reaching the wing.

The Hatfield Polytechnic Computer Centre

The DEC System 10 Computer of the Hatfield Polytechnic. The bottom photograph shows the drum plotter used to draw the aerofoils from the computed data for each section. *By kind permission of Hatfield Polytechnic.*

Aerofoil Sections

There is a descriptive index to the aerofoil sheets at the end of this book (commencing on page 205). The sheets themselves are separately numbered—from 1 to 83 in large bold figures – and these sheet numbers should not be confused with the *page* numbers elsewhere.

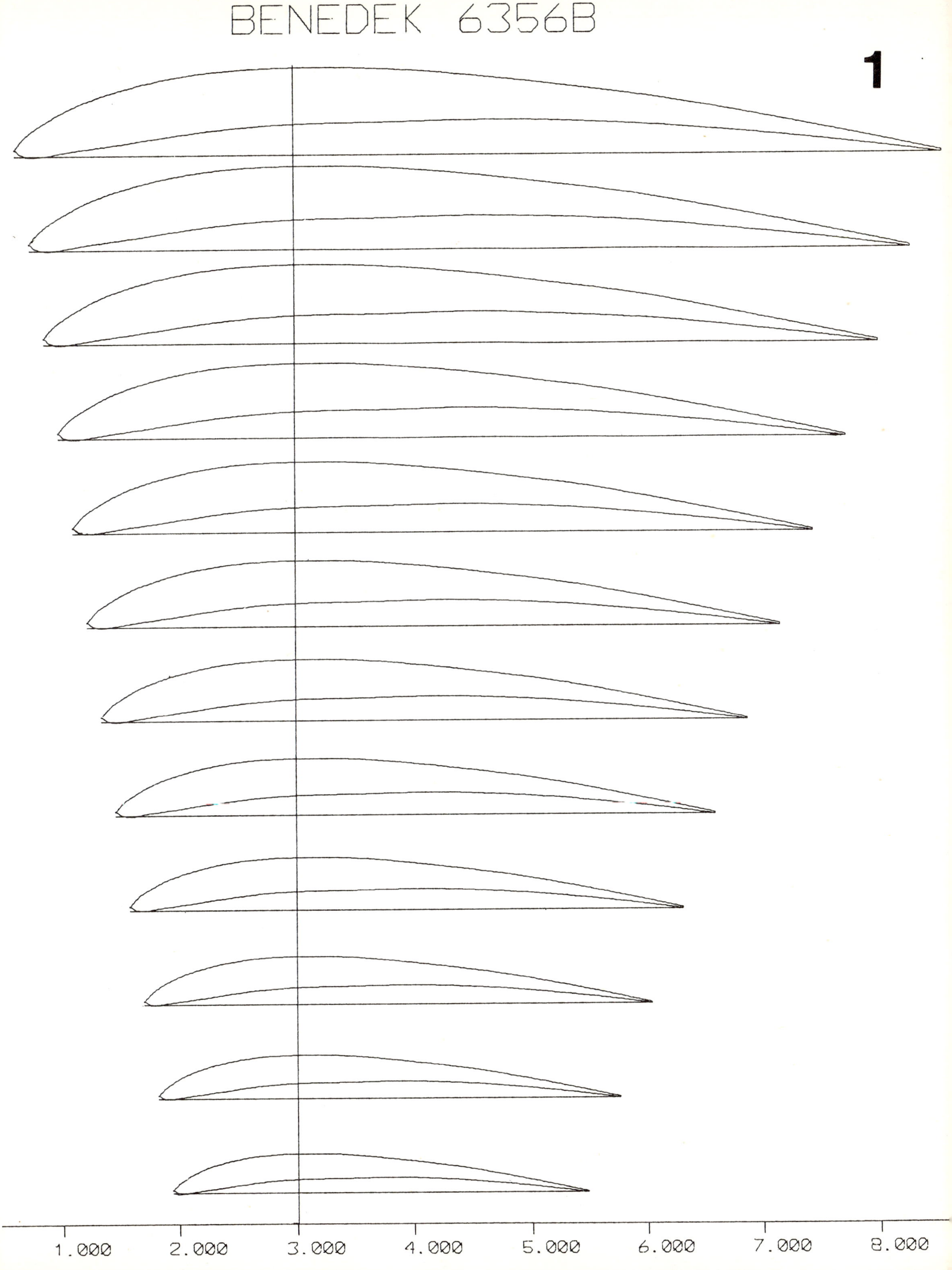
BENEDEK 6356B
1
1.000
2.000
3.000
4.000
5.000
6.000
7.000
8.000

BENEDEK 6356B, SKIMMED

2

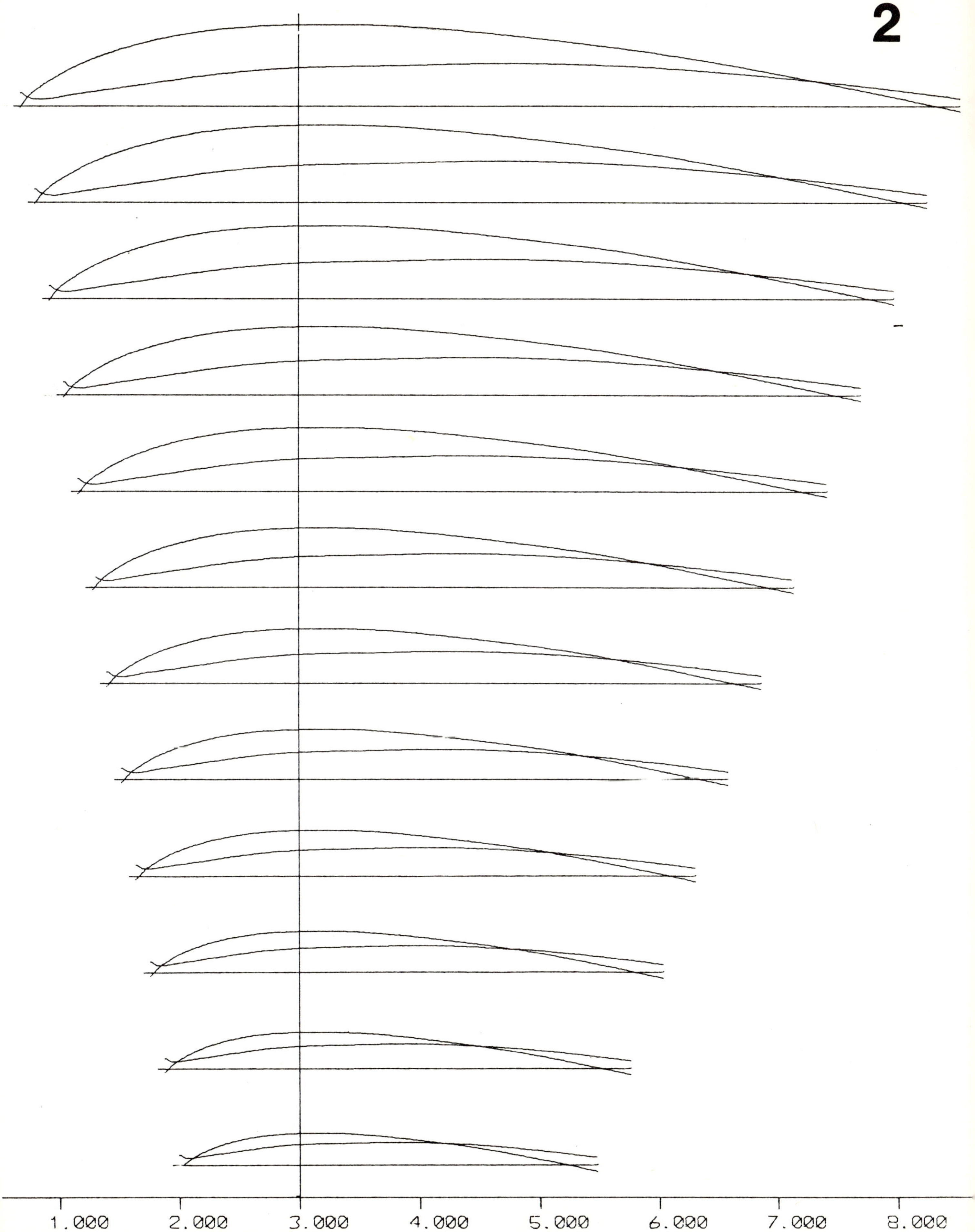

BENEDEK 6407E

3

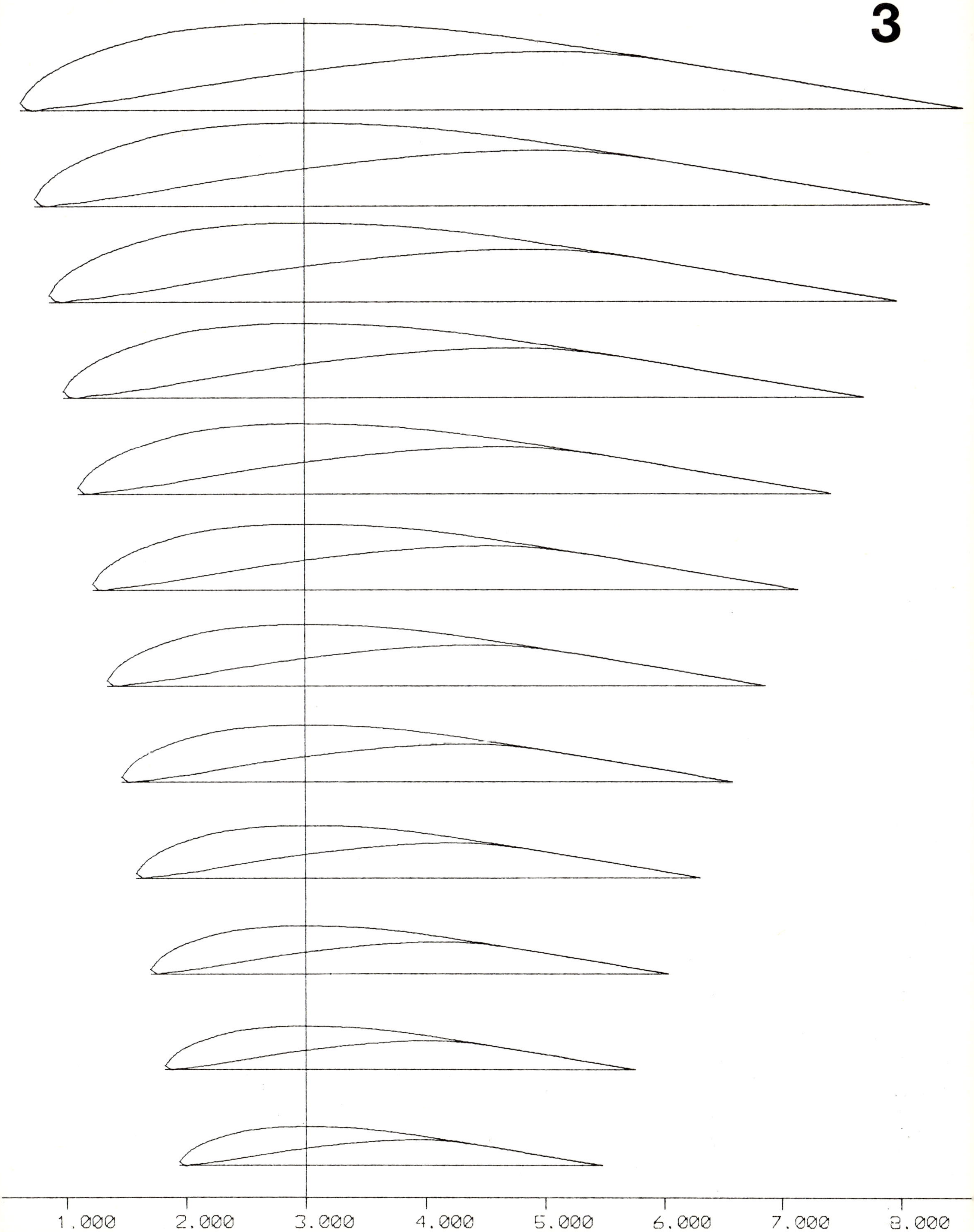

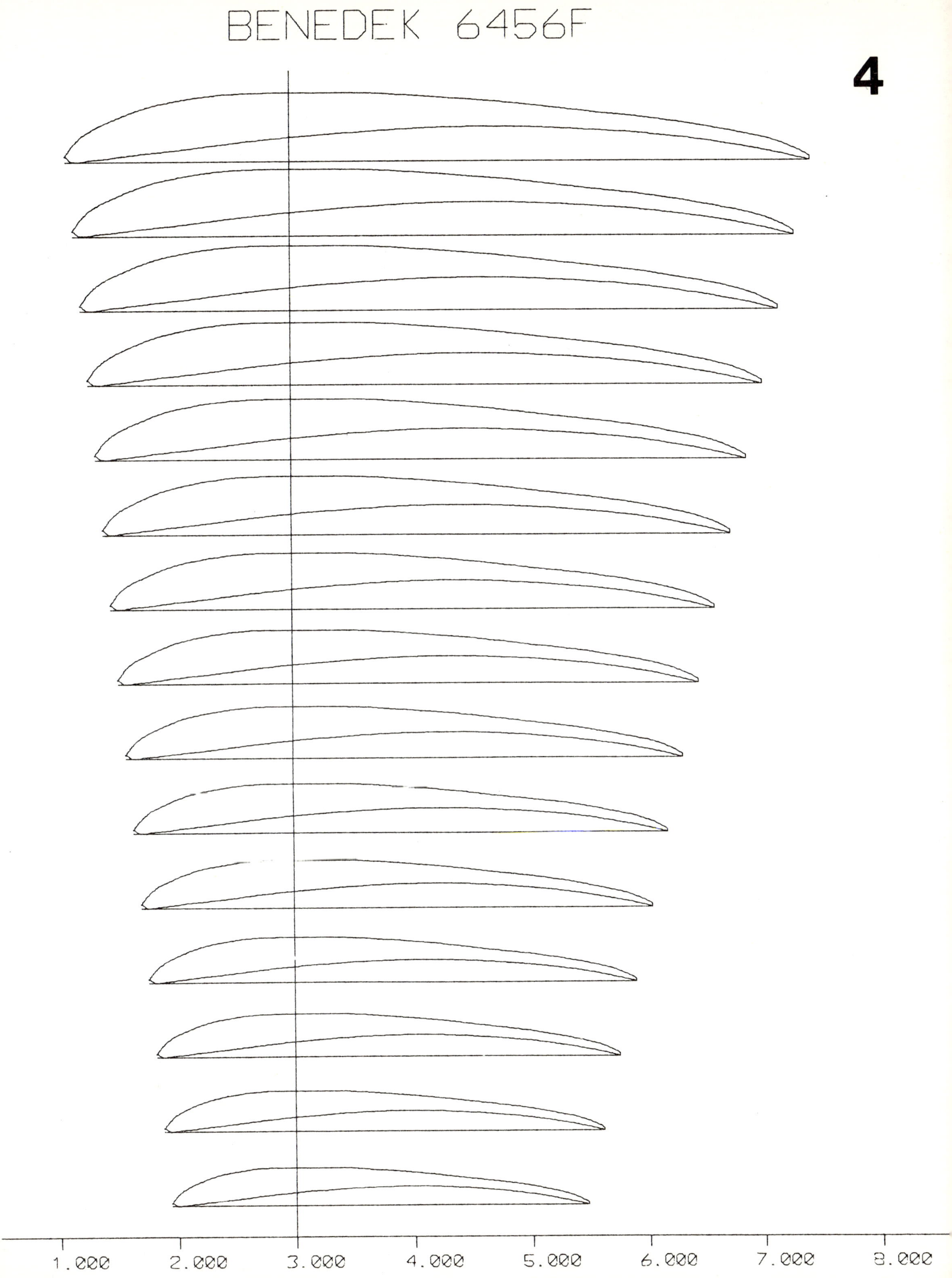
BENEDEK 6456F
4
1.000
2.000
3.000
4.000
5.000
6.000
7.000
8.000

BENEDEK 7406F

5

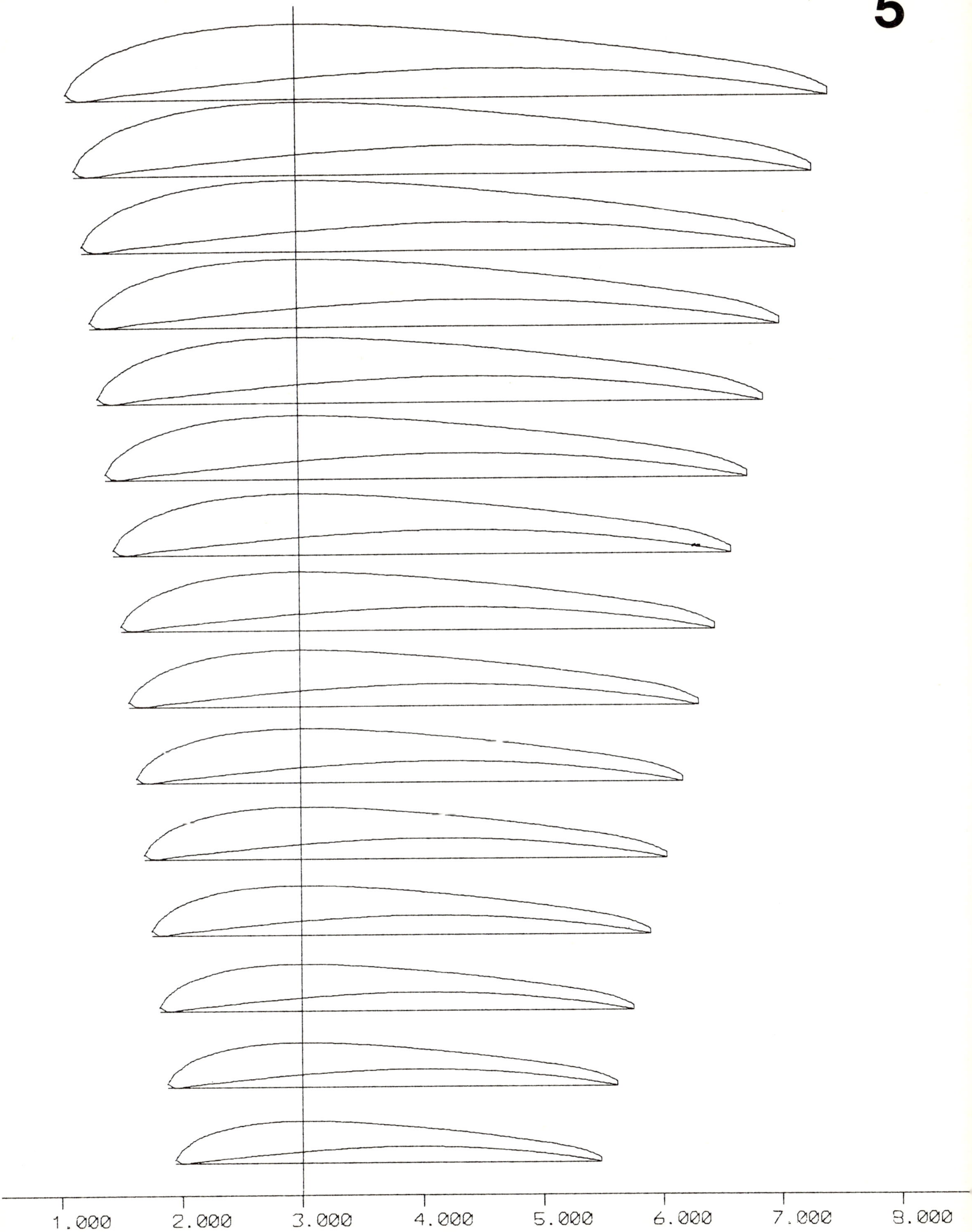

BENEDEK 8356B3

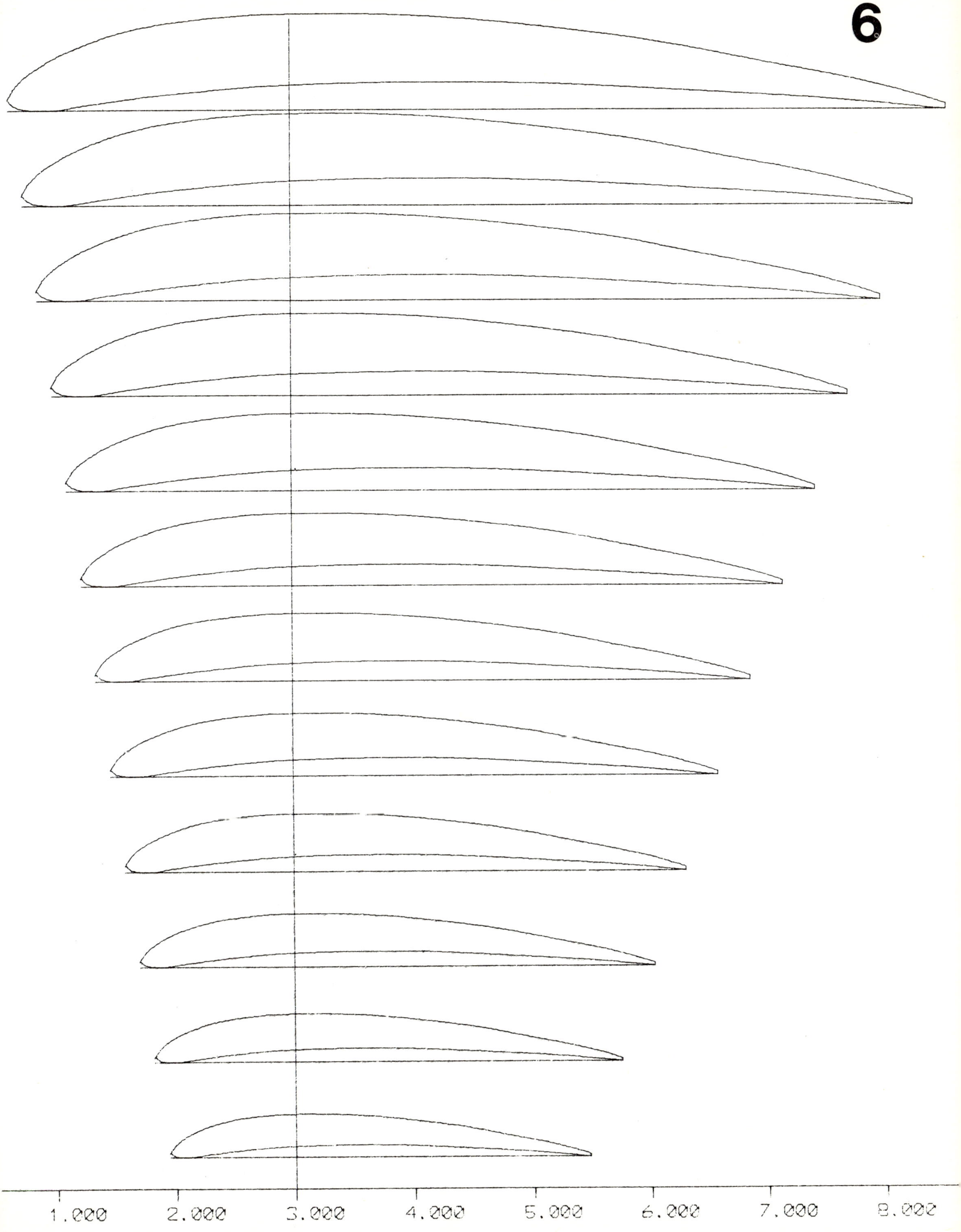

BENEDEK 8535B2

7

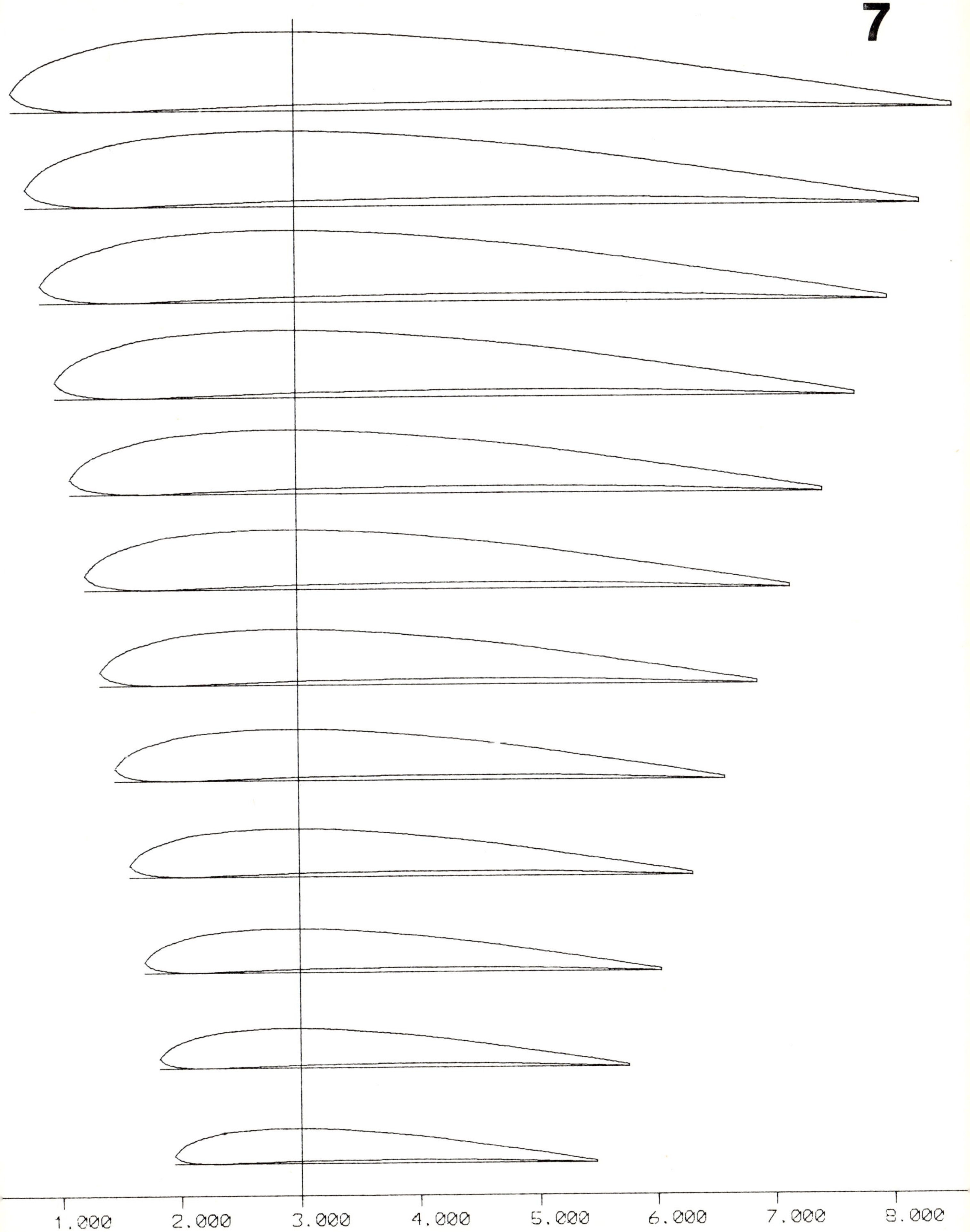

BENEDEK 8535B2, SKIMMED

8

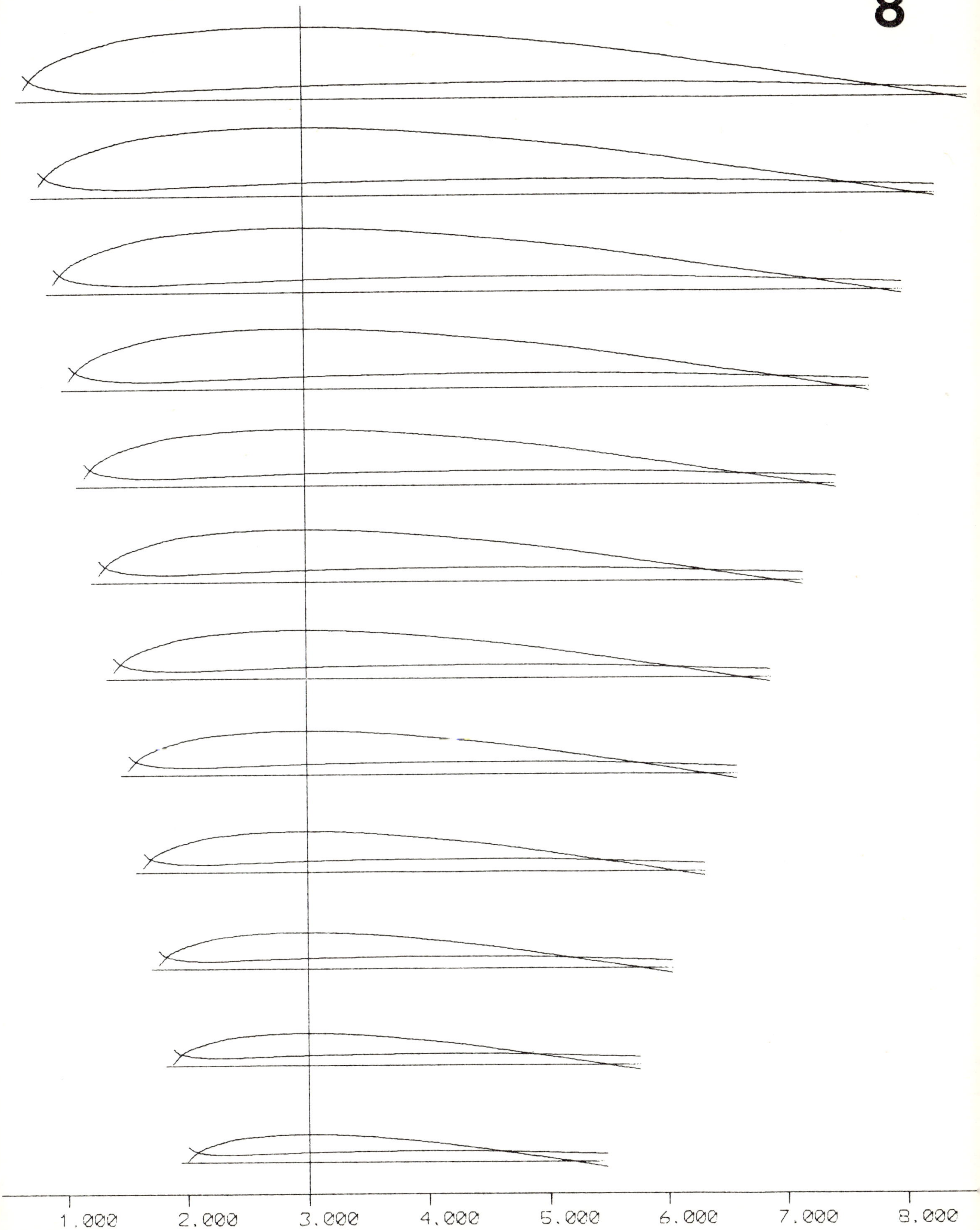

BENEDEK 8556B

9

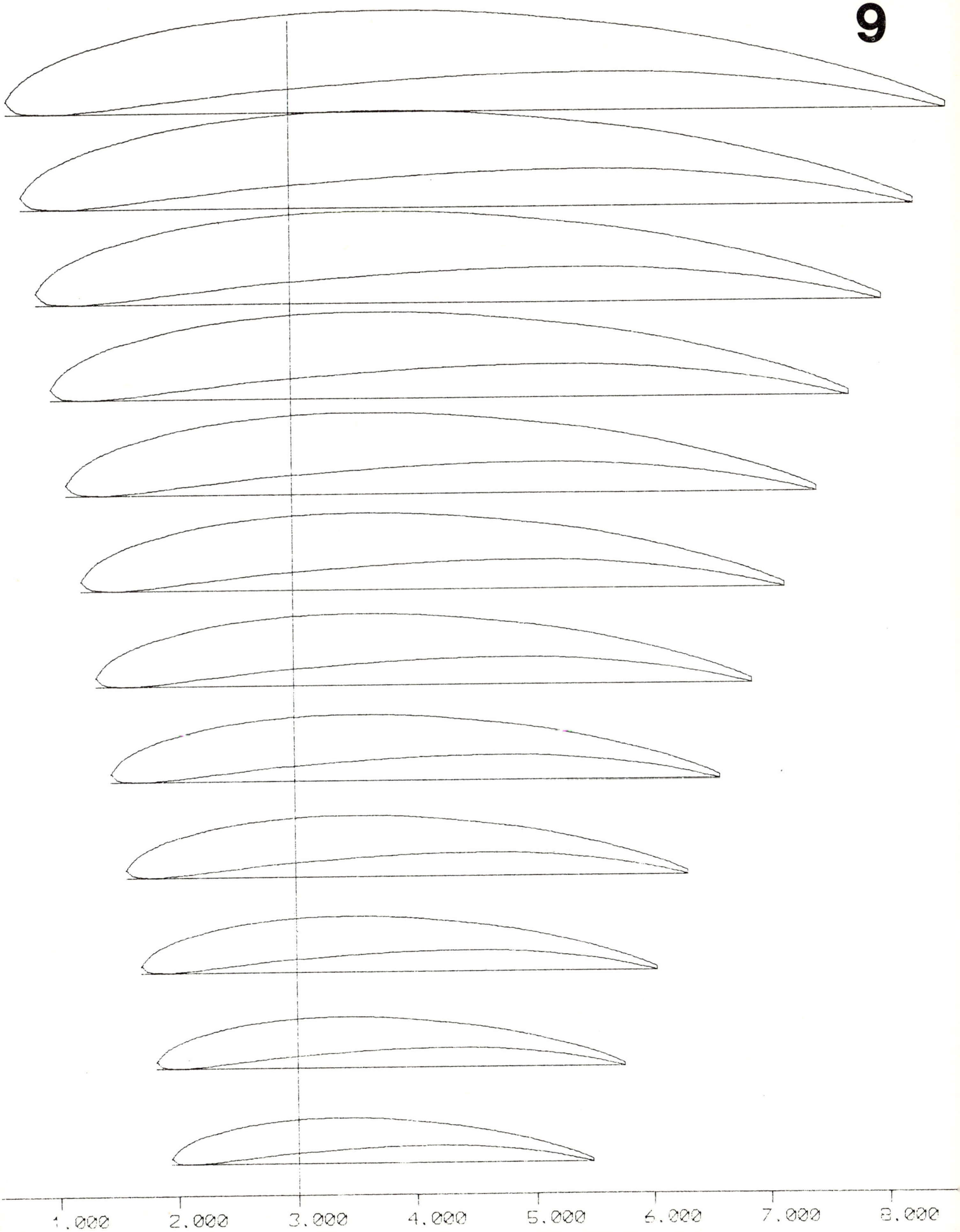

BENEDEK 8556B, SKIMMED

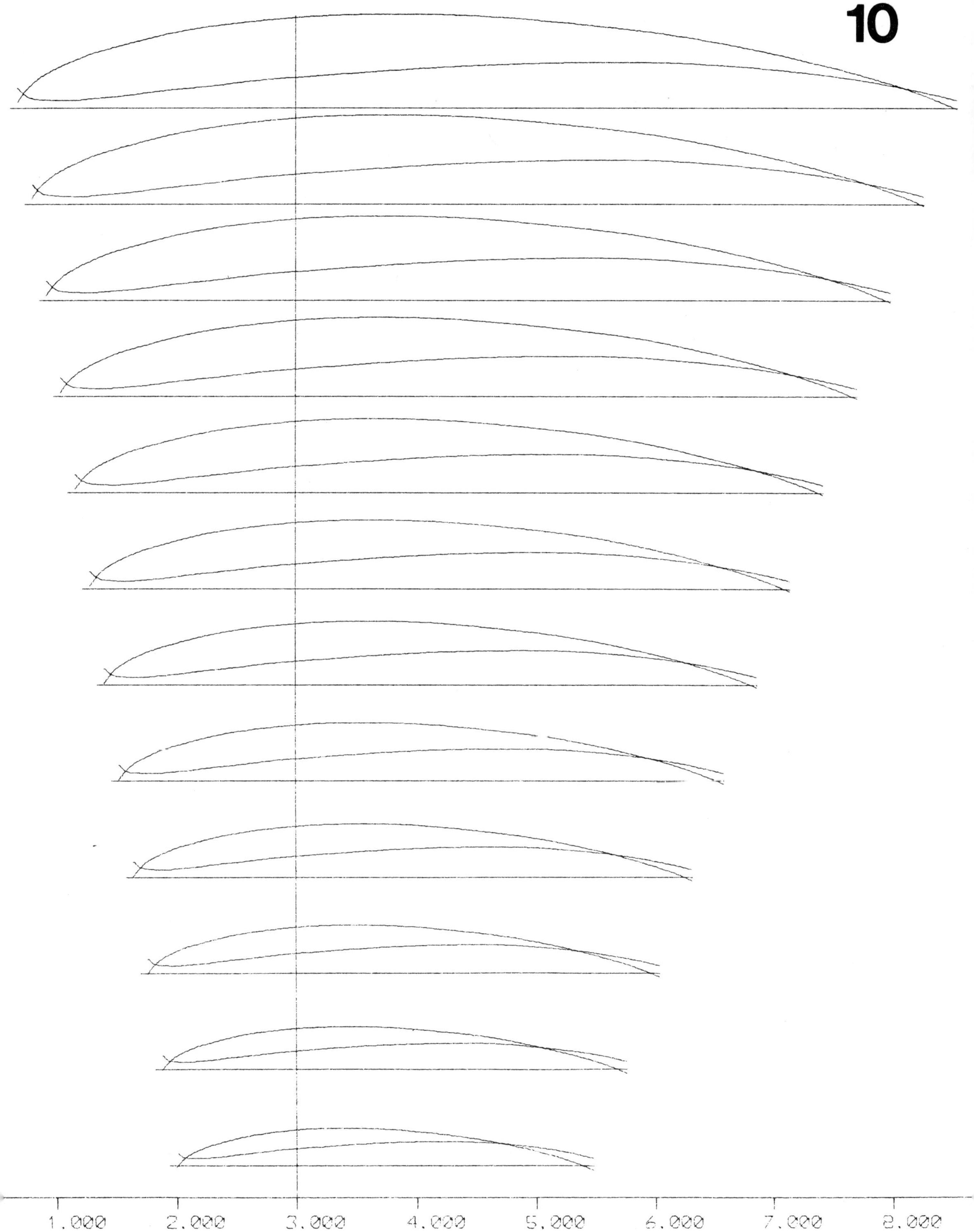

CLARK Y

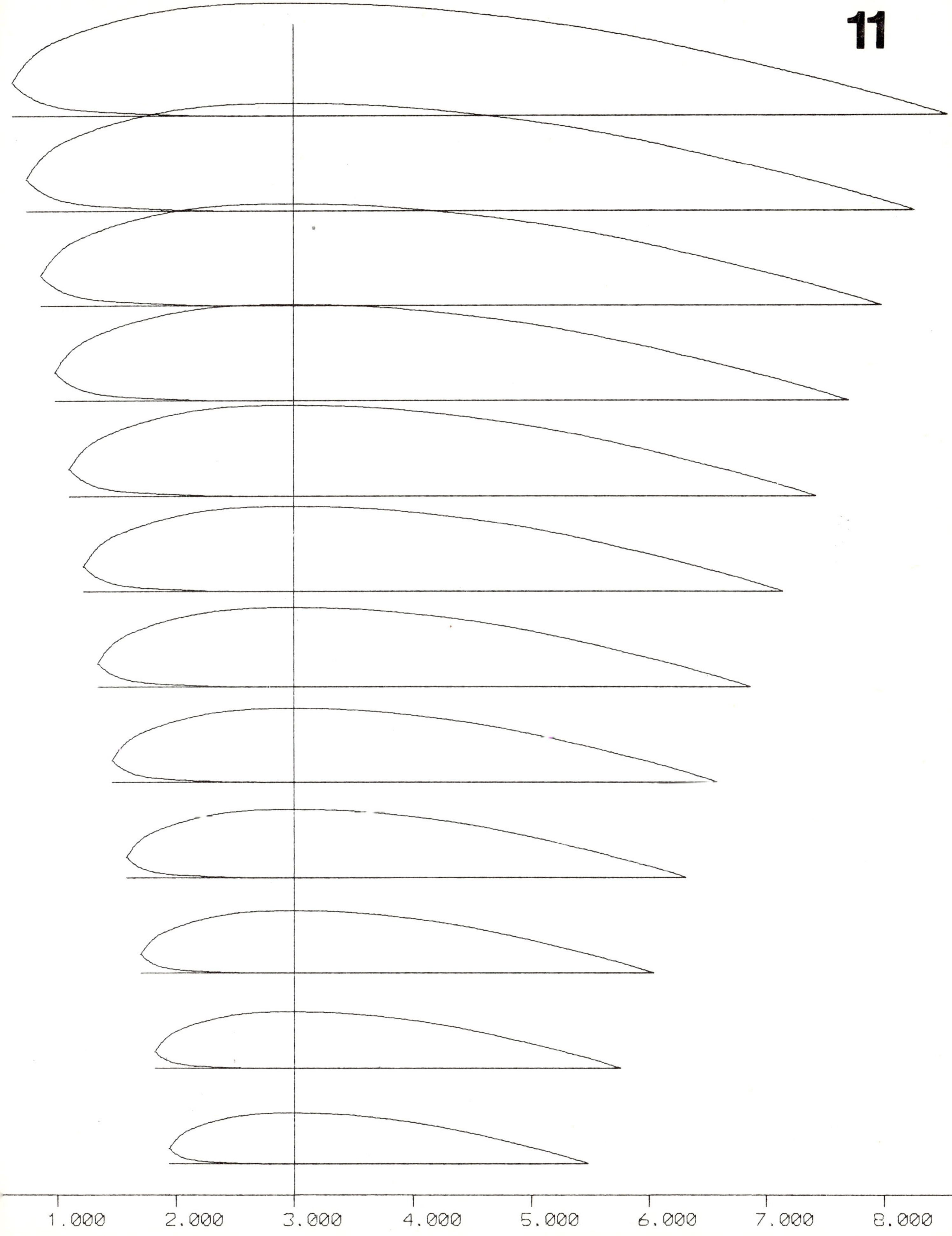

CLARK Y, SKIMMED

12

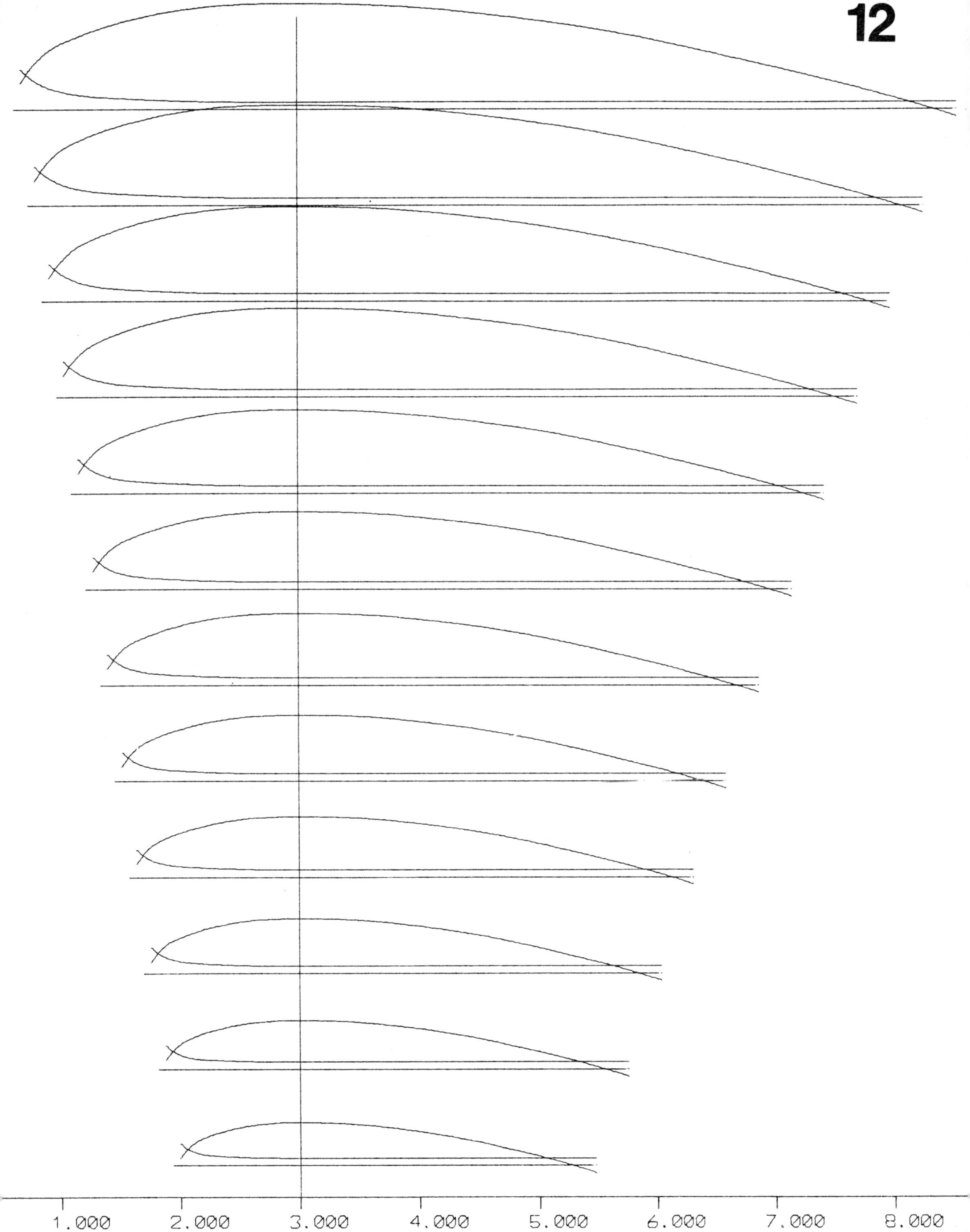

CZEPA

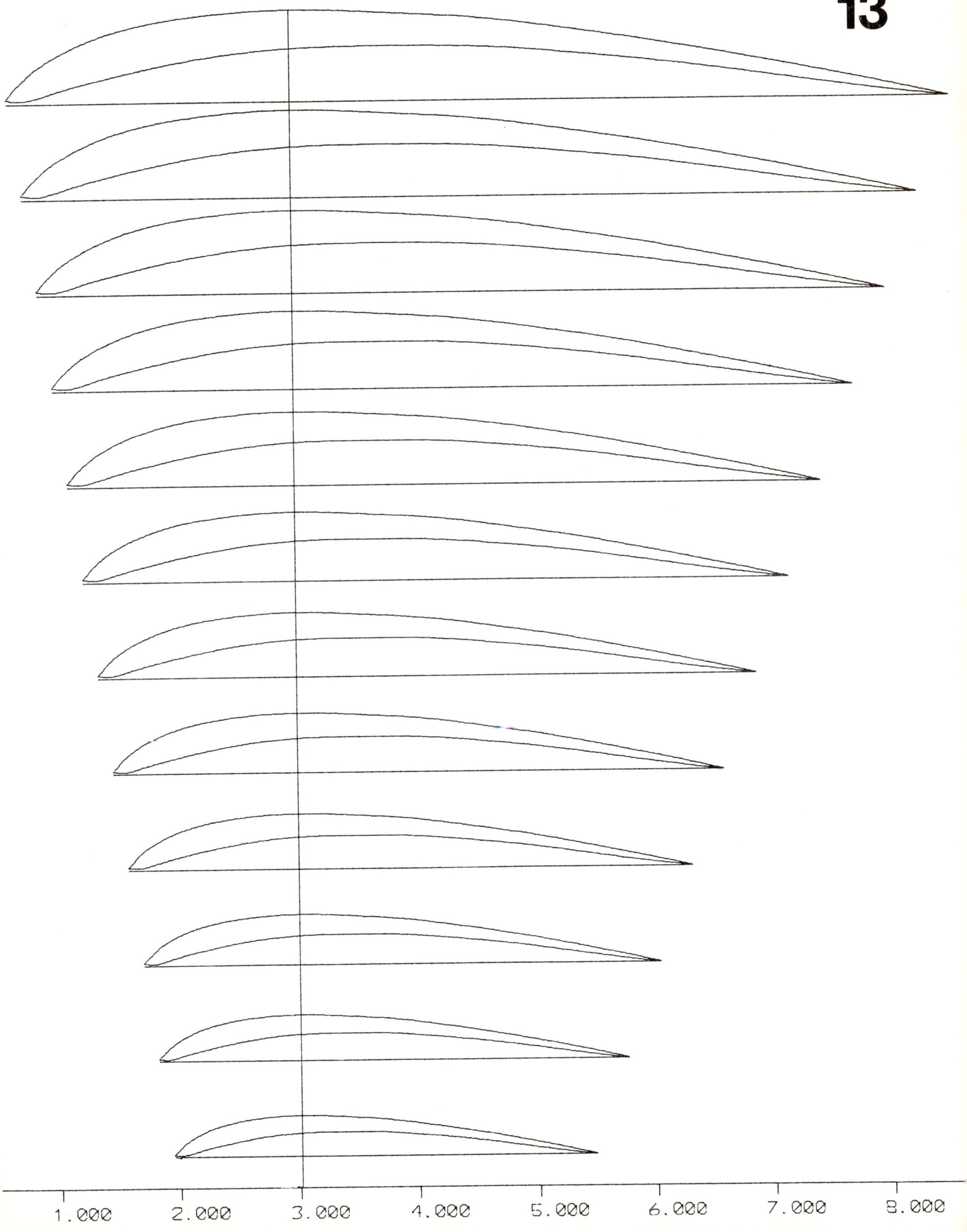

DAVIS A.93,B.17

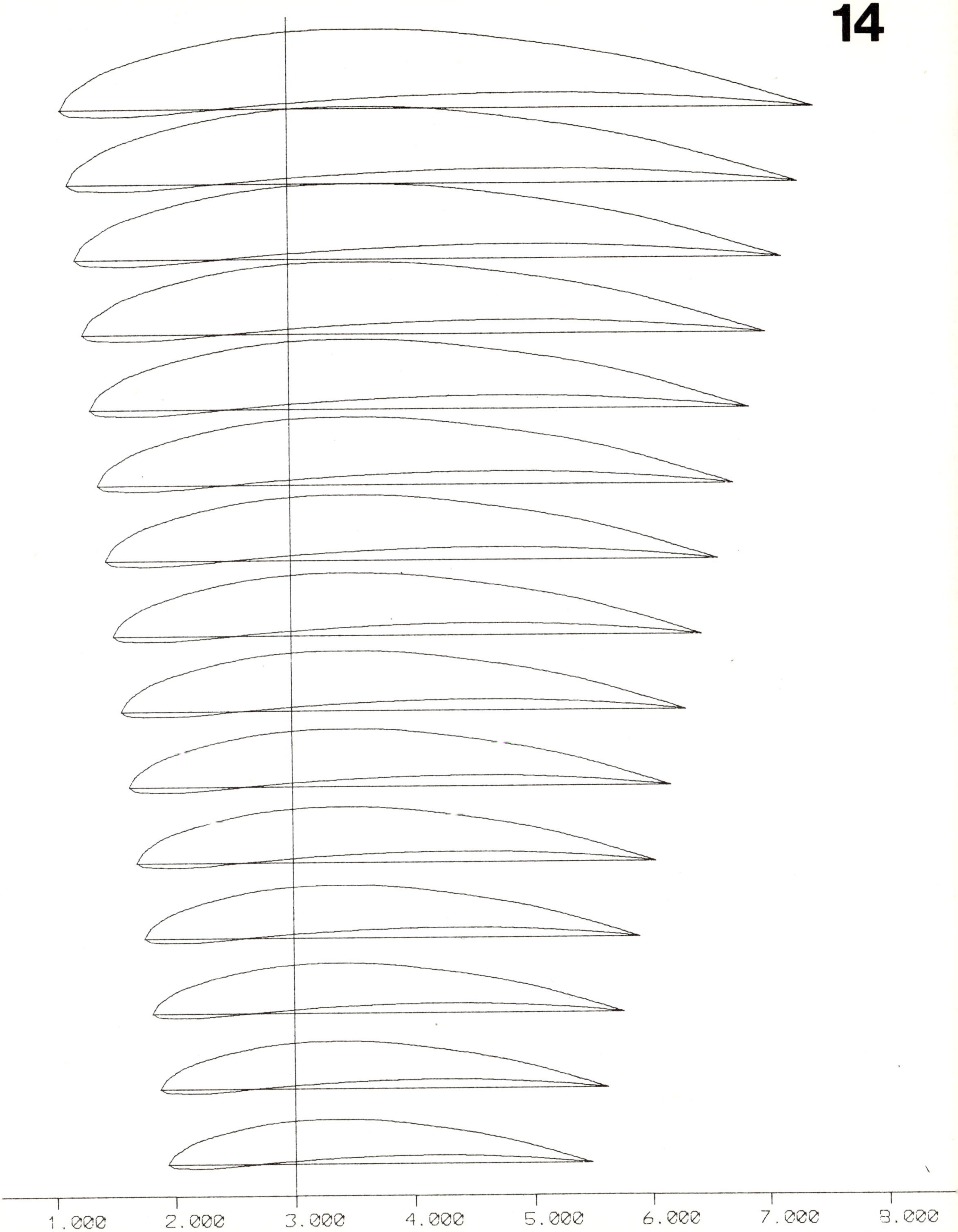

15

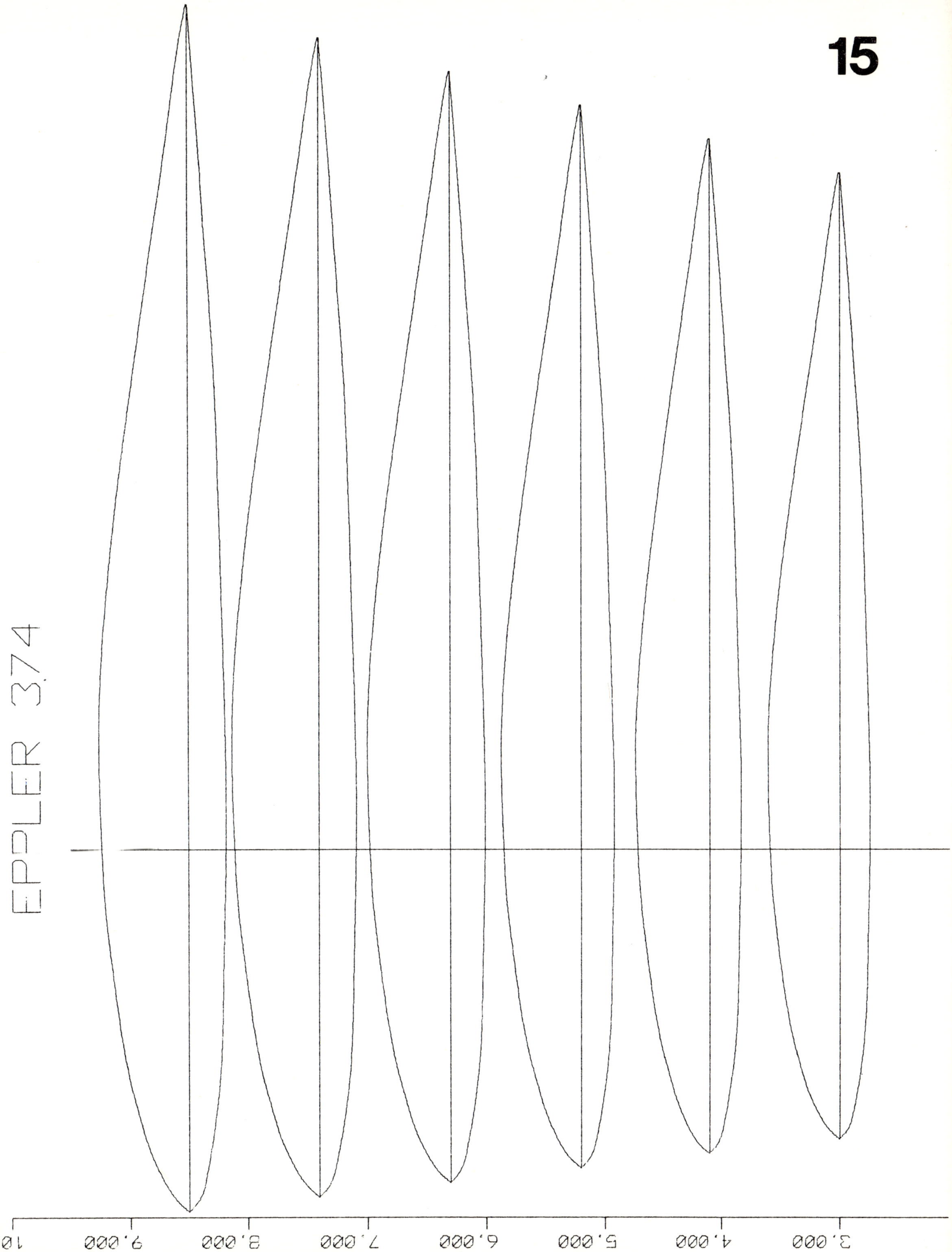

EPPLER 374, SKIMMED

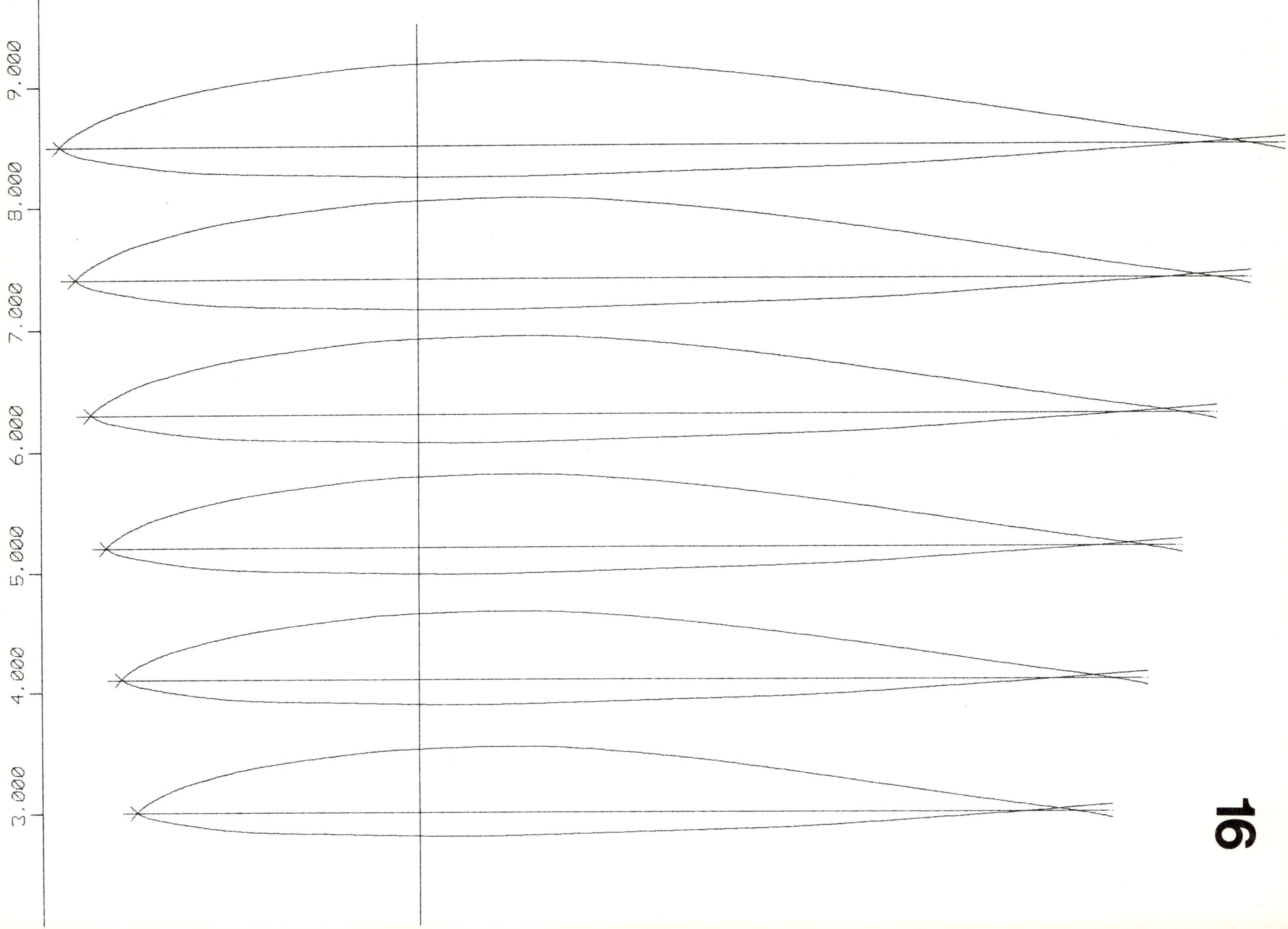

16

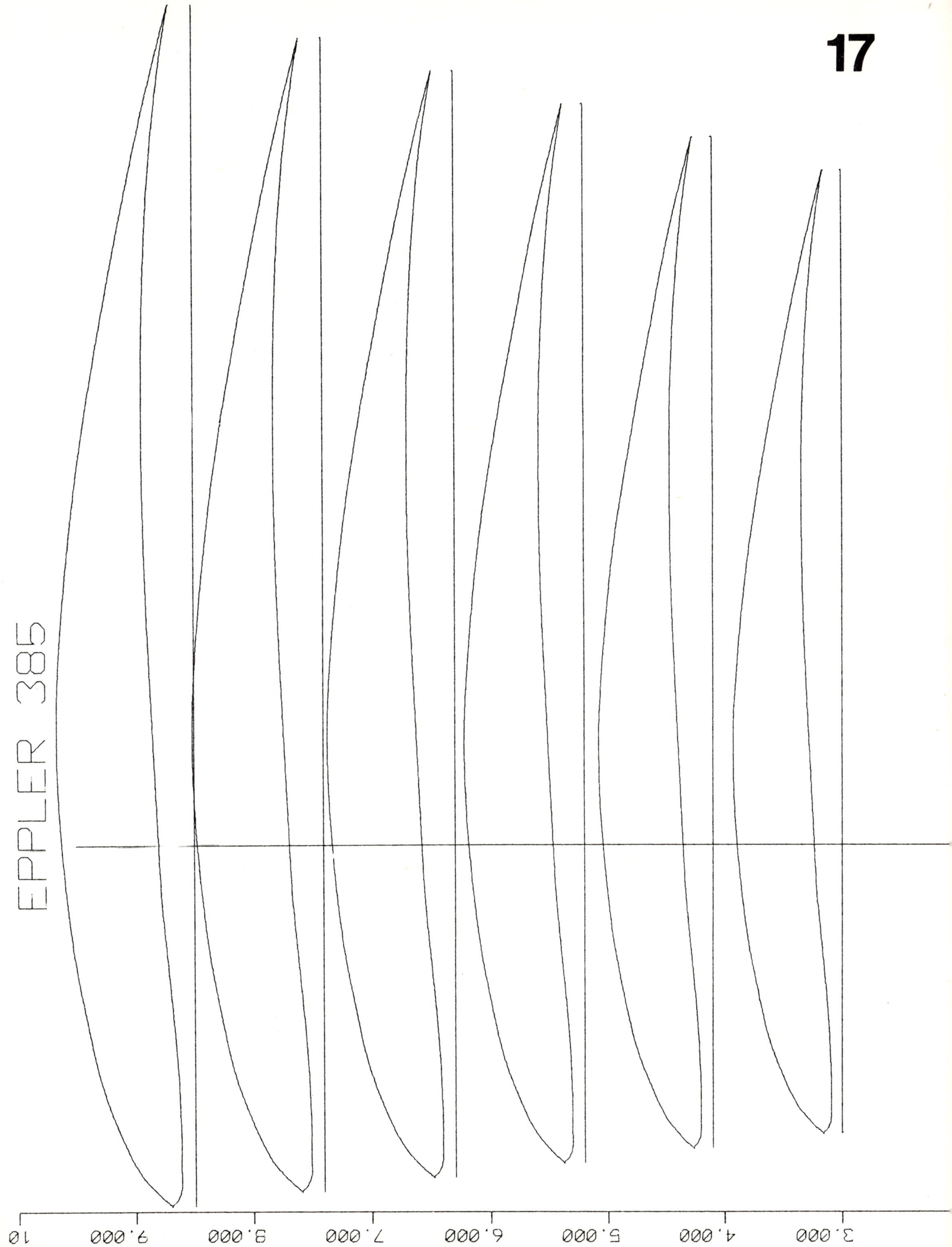
EPPLER 385
10
9.000
8.000
7.000
6.000
5.000
4.000
3.000

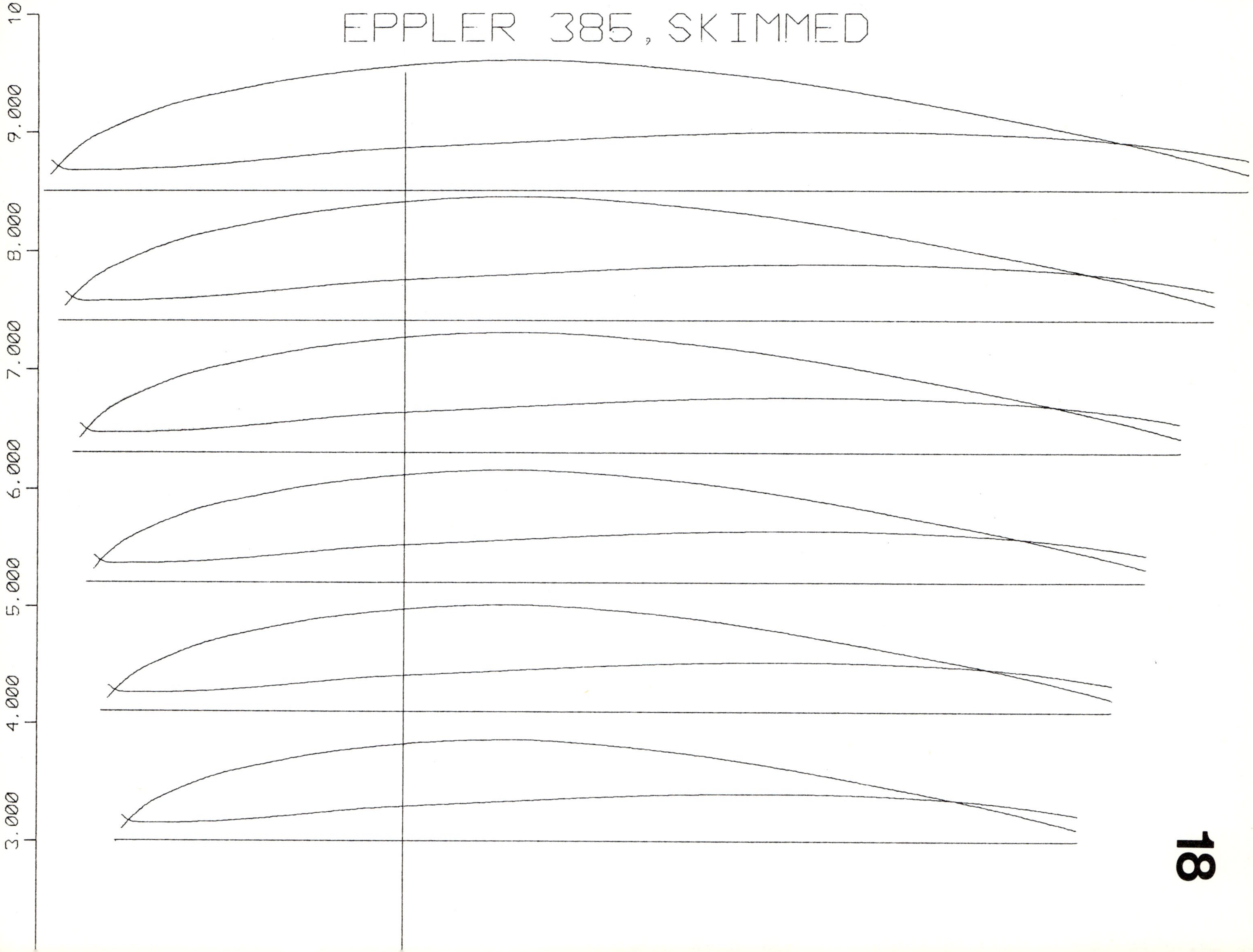

18

EPPLER 385-387 BLENDED

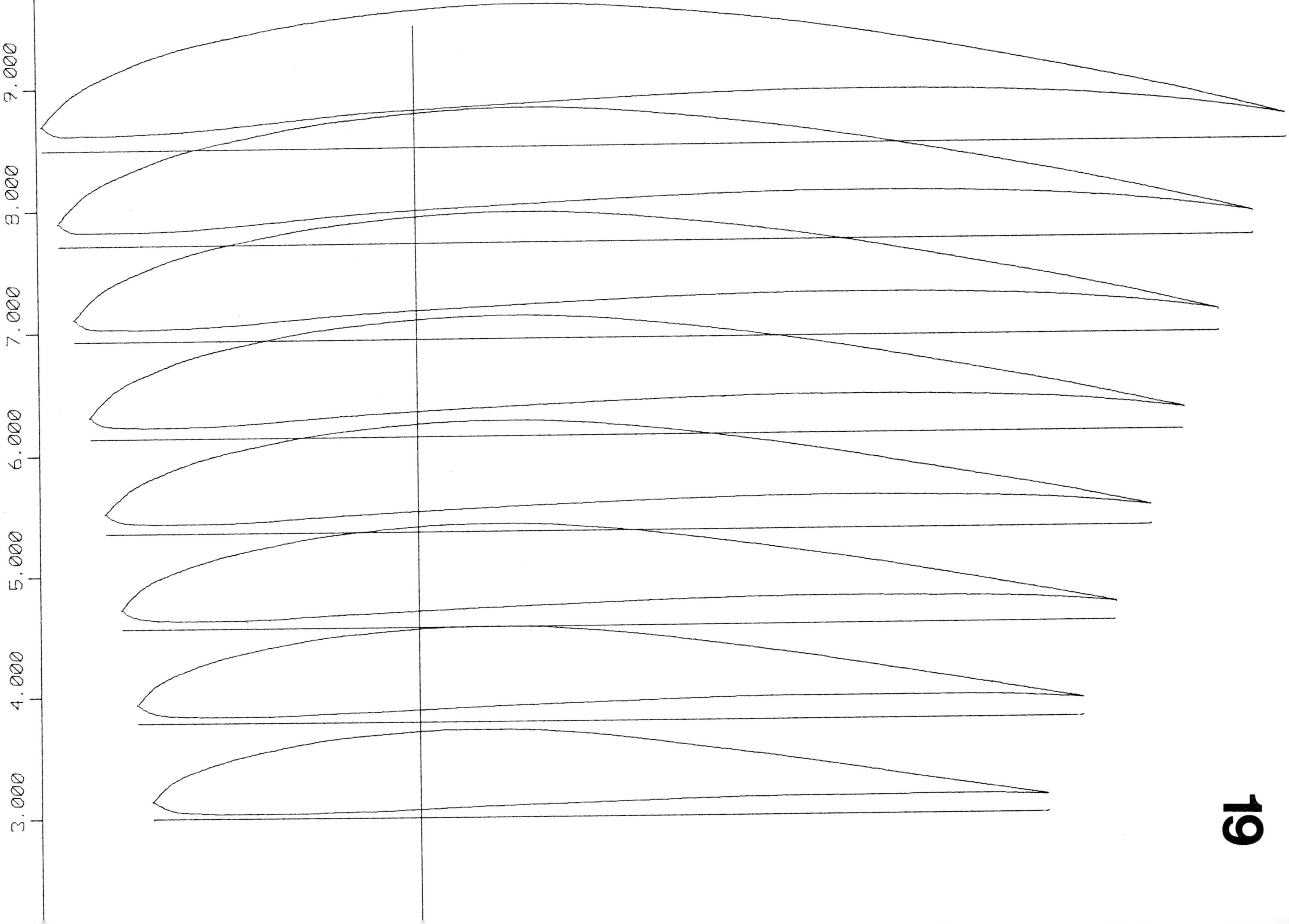

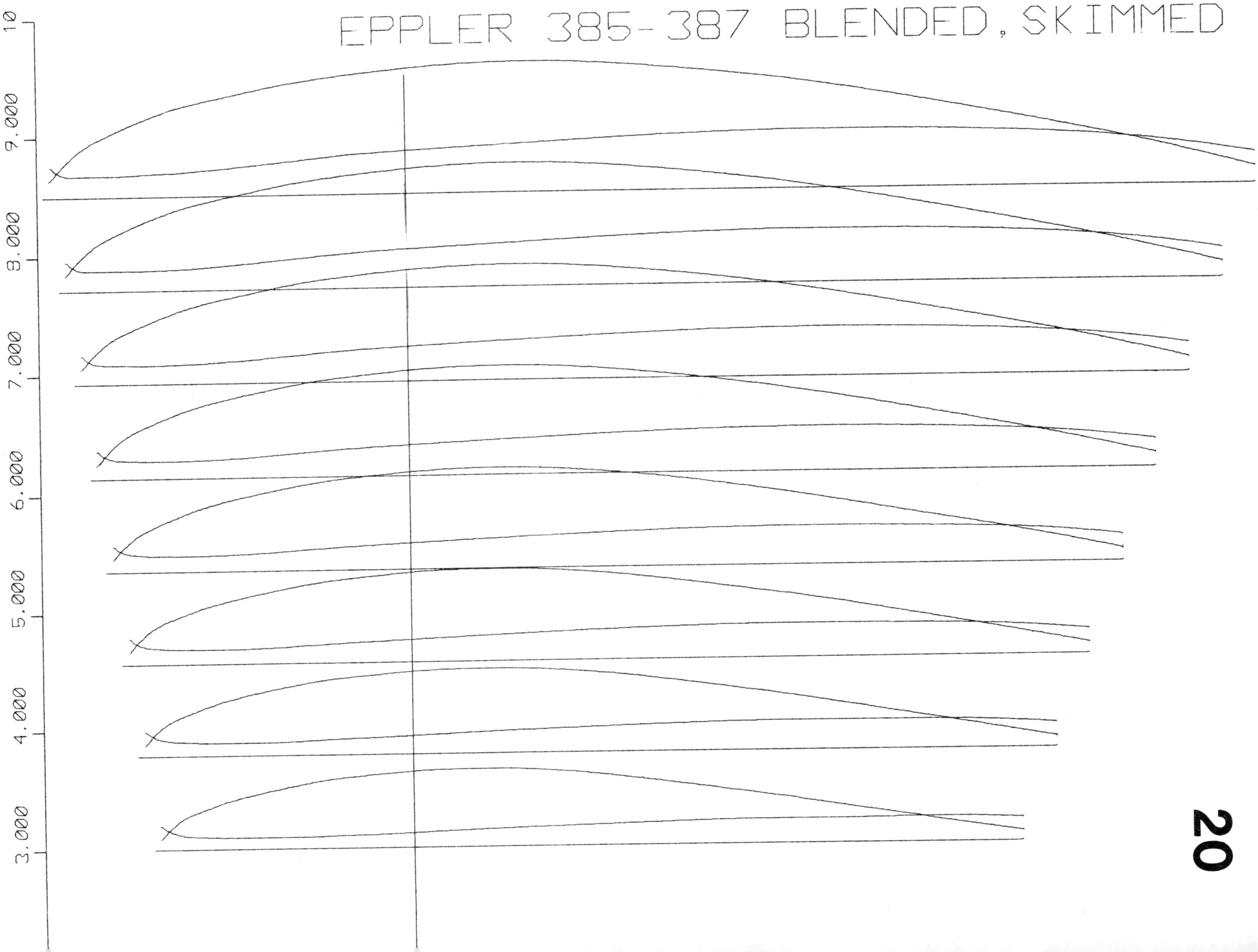
EPPLER 385-387 BLENDED, SKIMMED
10
9.000
8.000
7.000
6.000
5.000
4.000
3.000

20

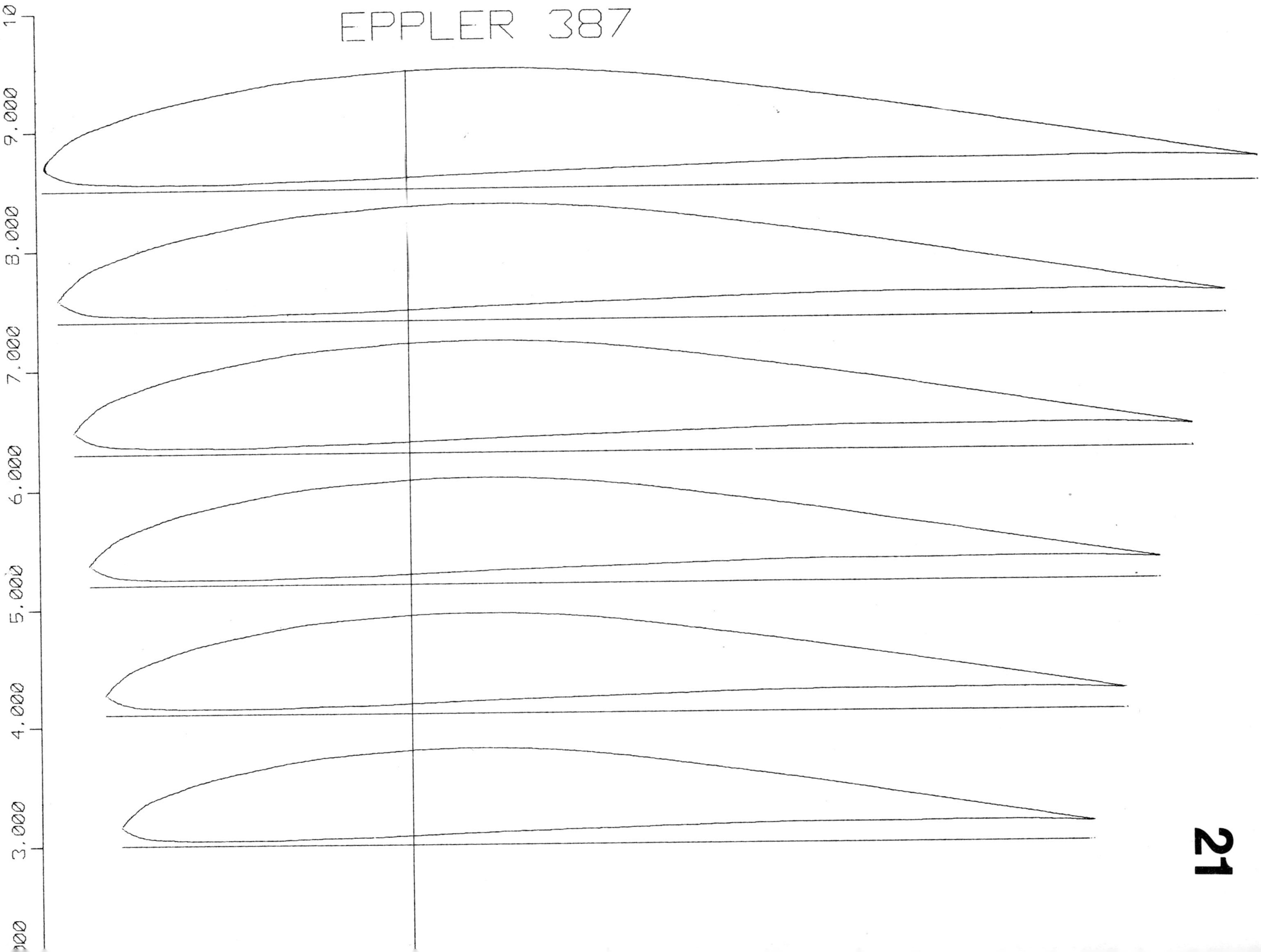
EPPLER 387
10
9.000
8.000
7.000
6.000
5.000
4.000
3.000

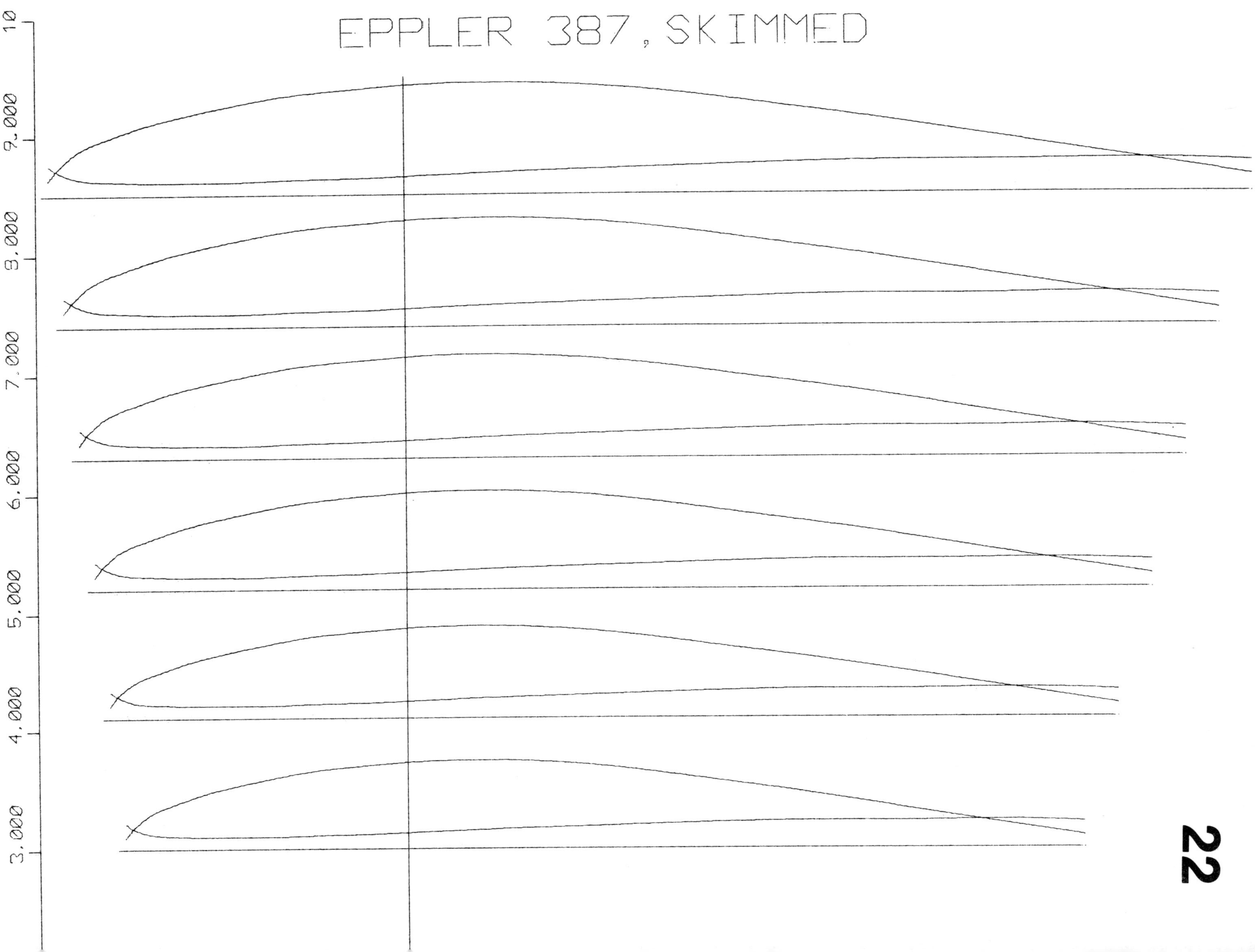
EPPLER 387, SKIMMED
10
9.000
8.000
7.000
6.000
5.000
4.000
3.000

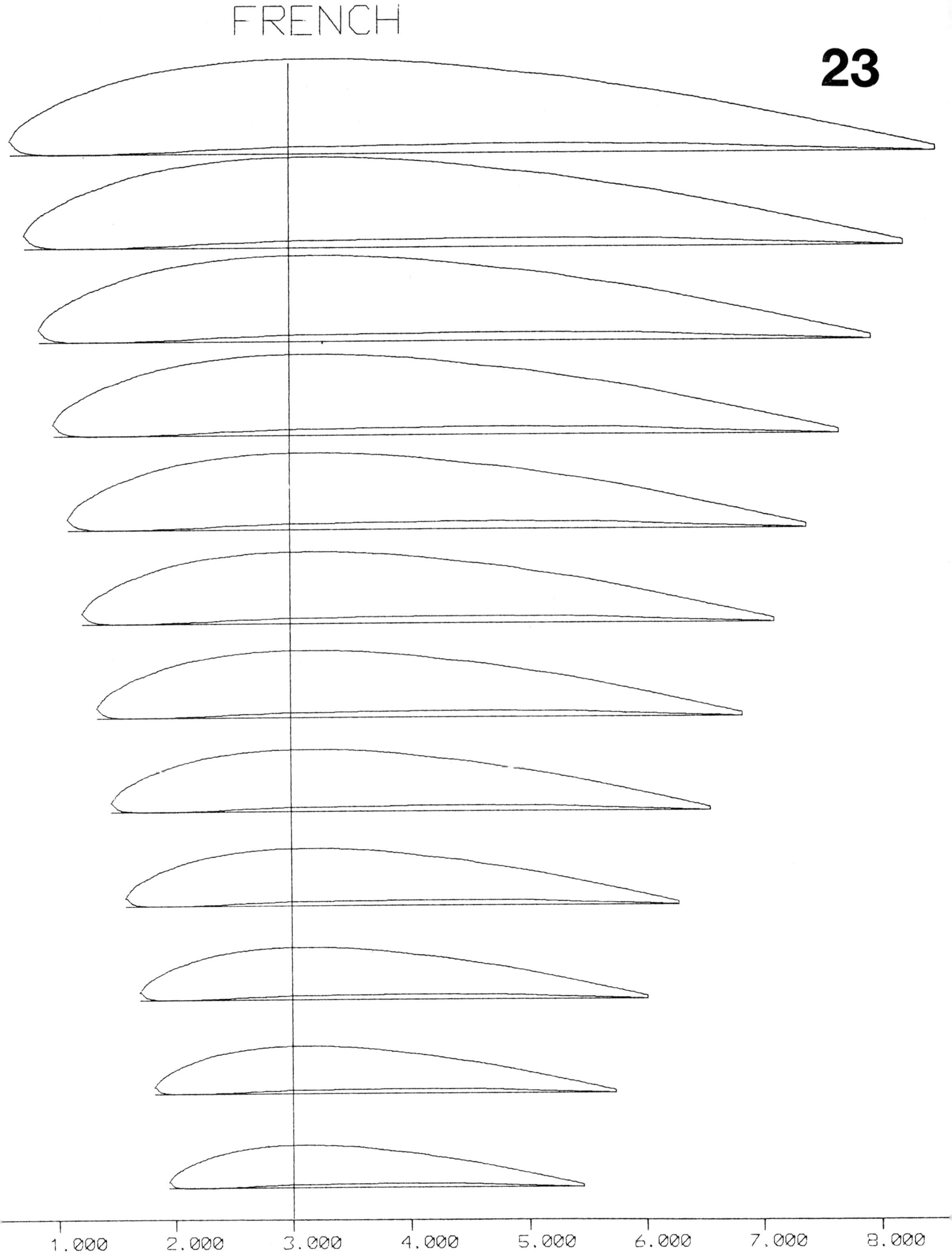
FRENCH
1.000
2.000
3.000
4.000
5.000
6.000
7.000
8.000

FRENCH, SKIMMED

24

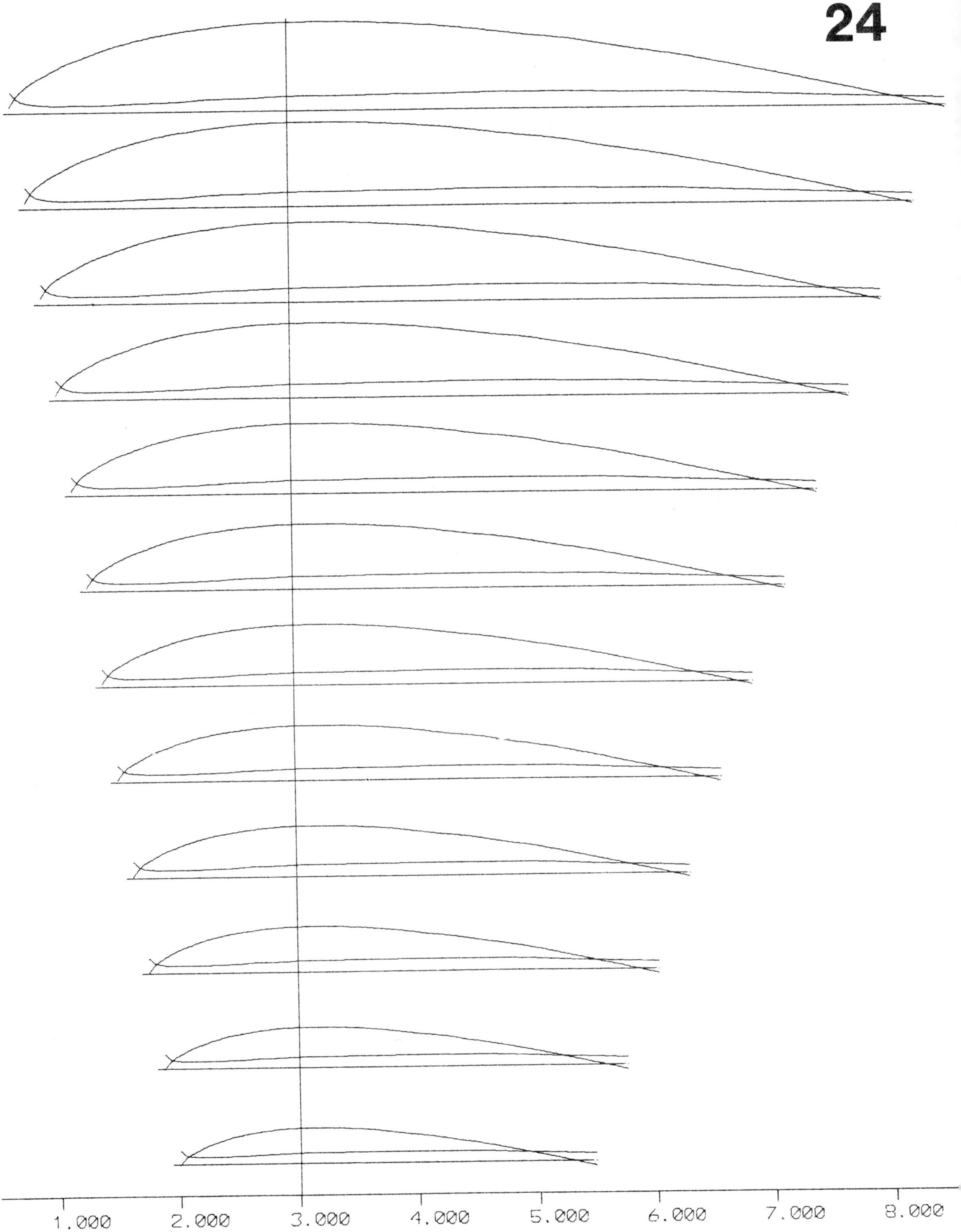

GOTTINGEN 417

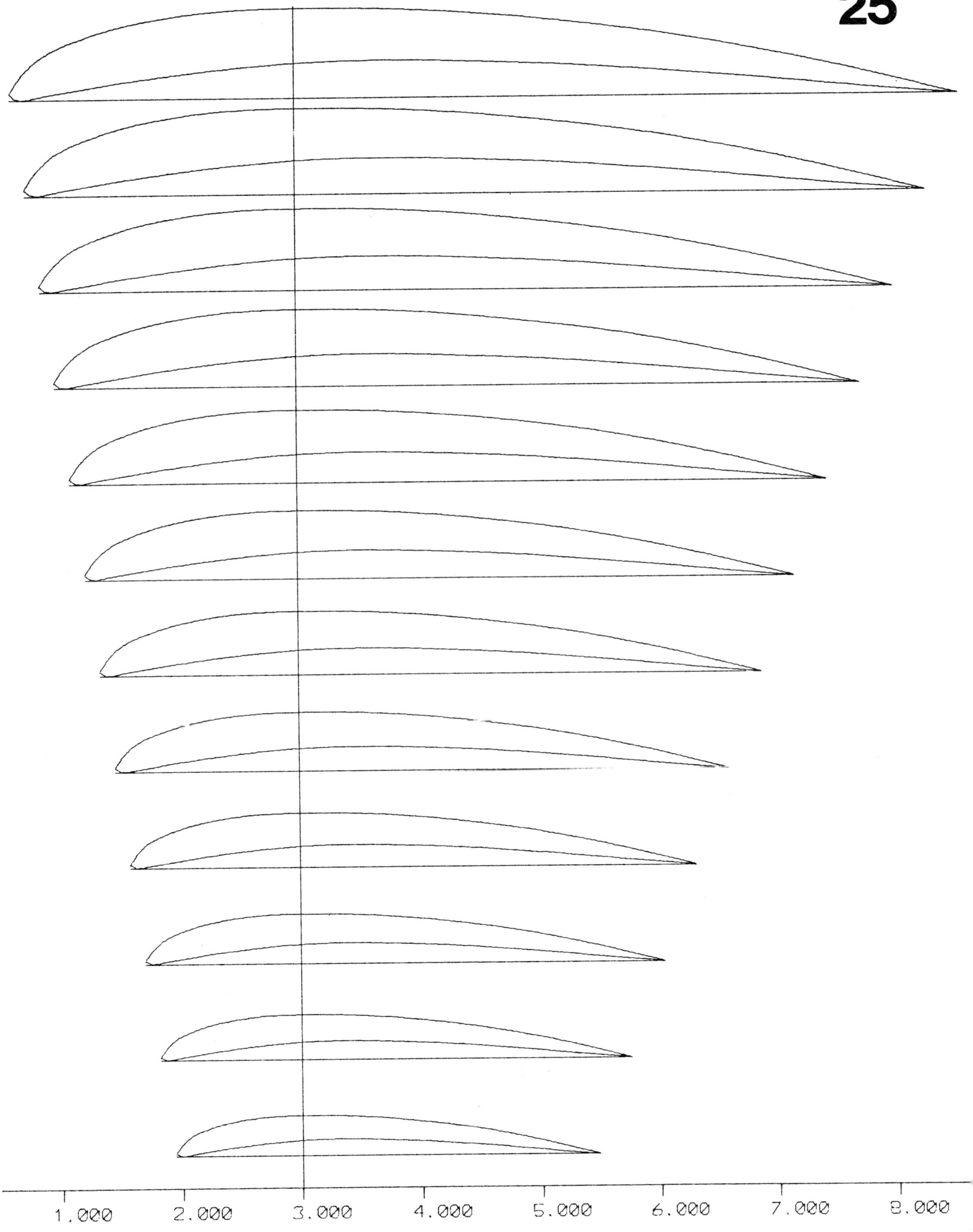

GOTTINGEN 495

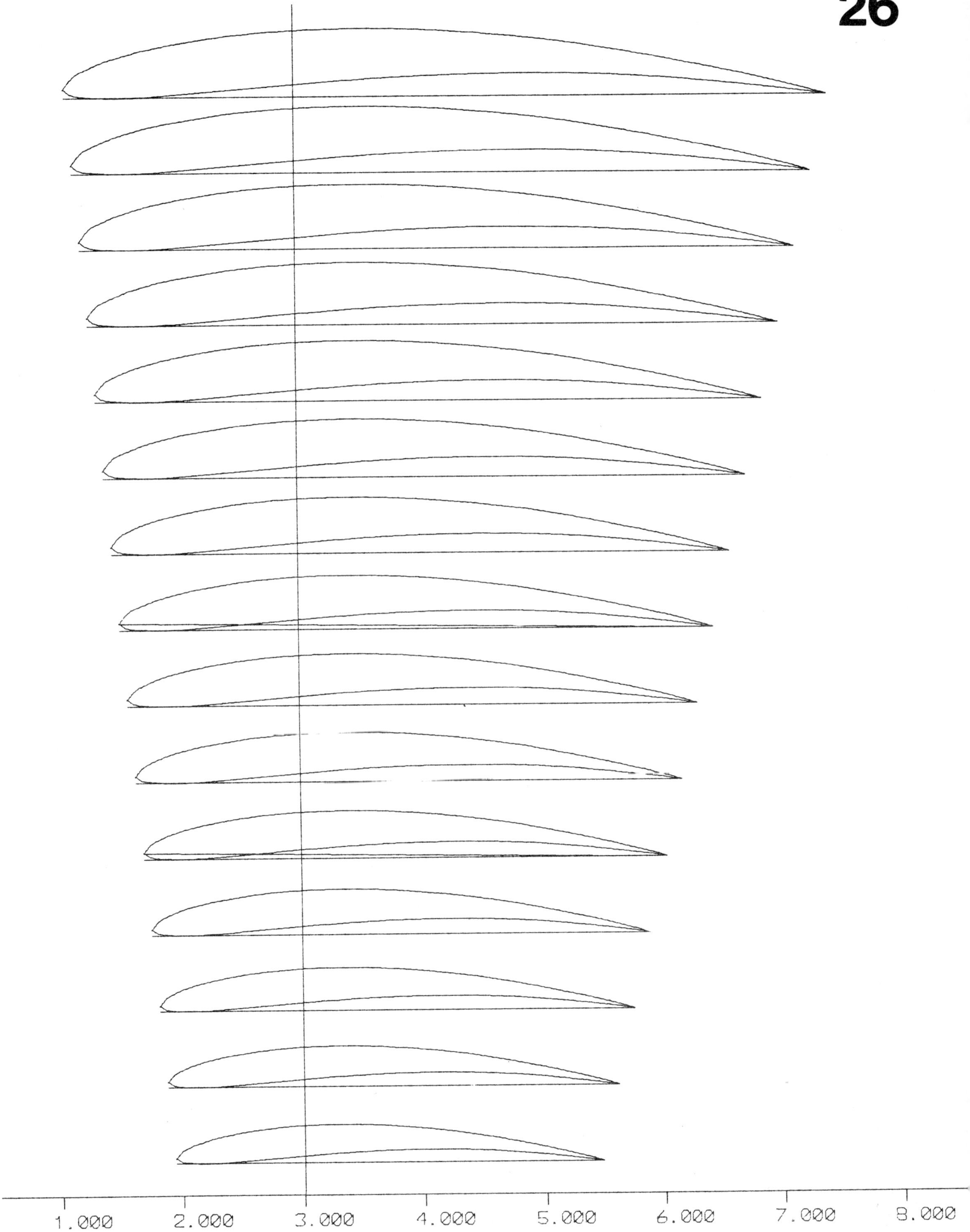

GOTTINGEN 6Ø2

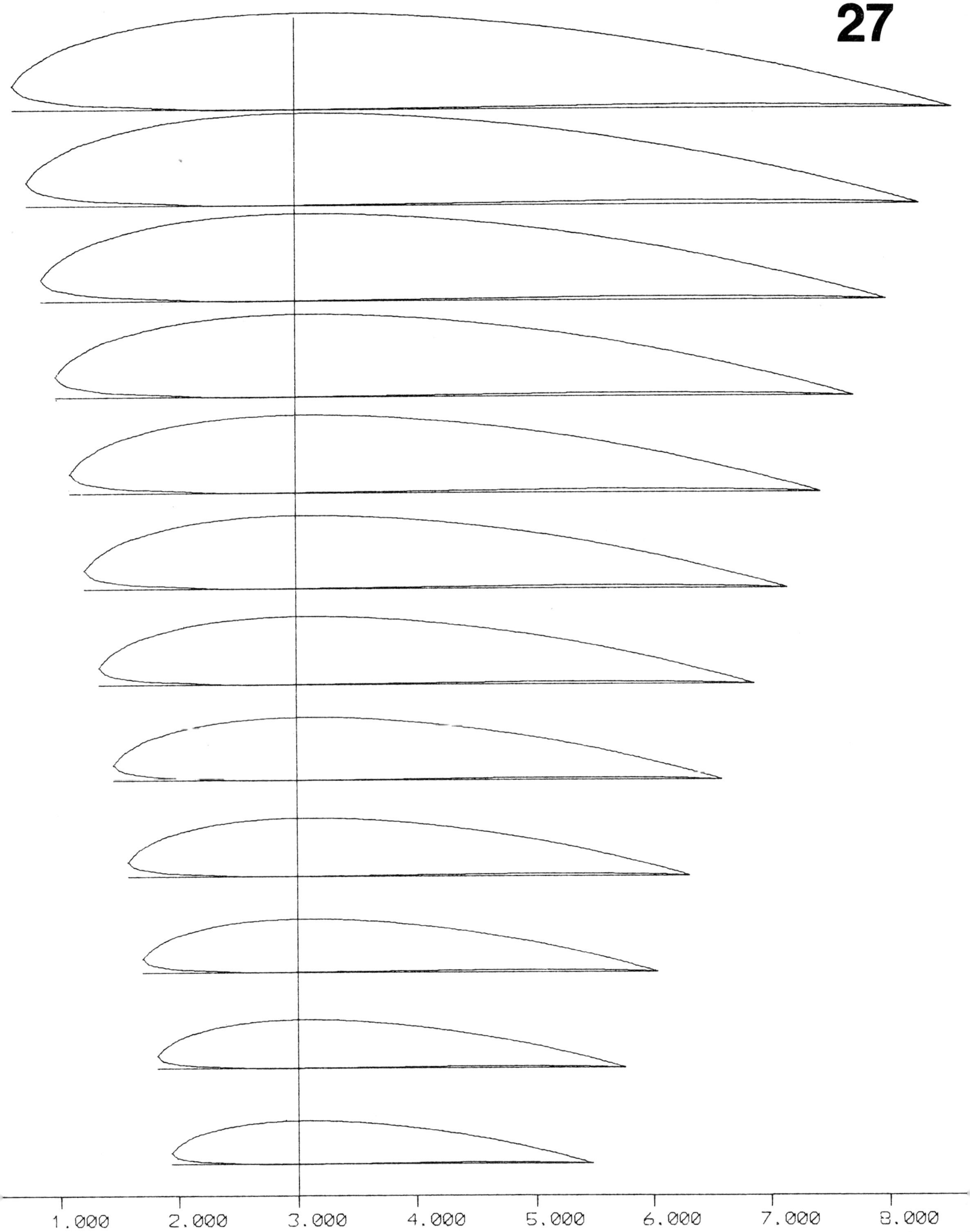

GOTTINGEN 602, SKIMMED

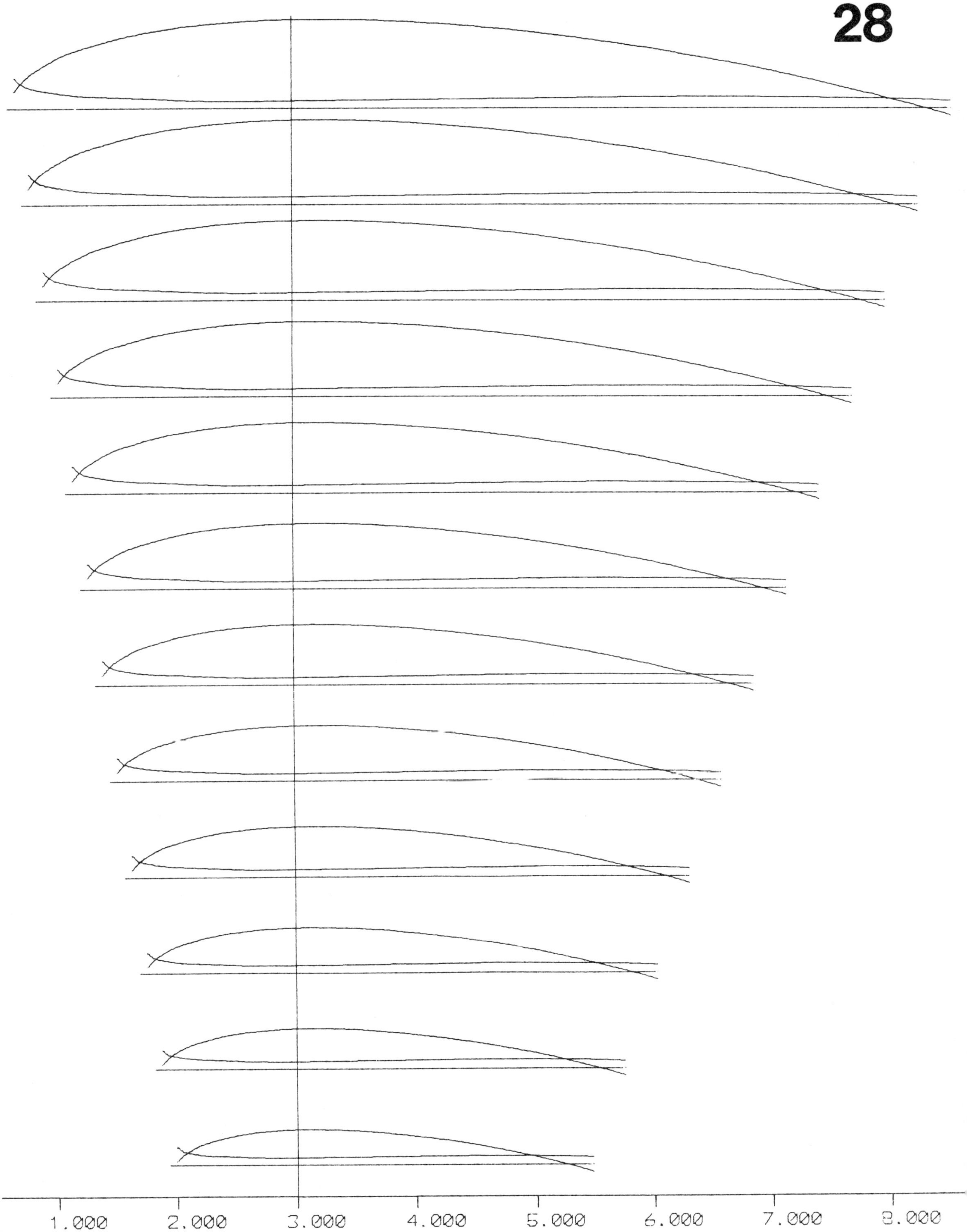

GOTTINGEN 795

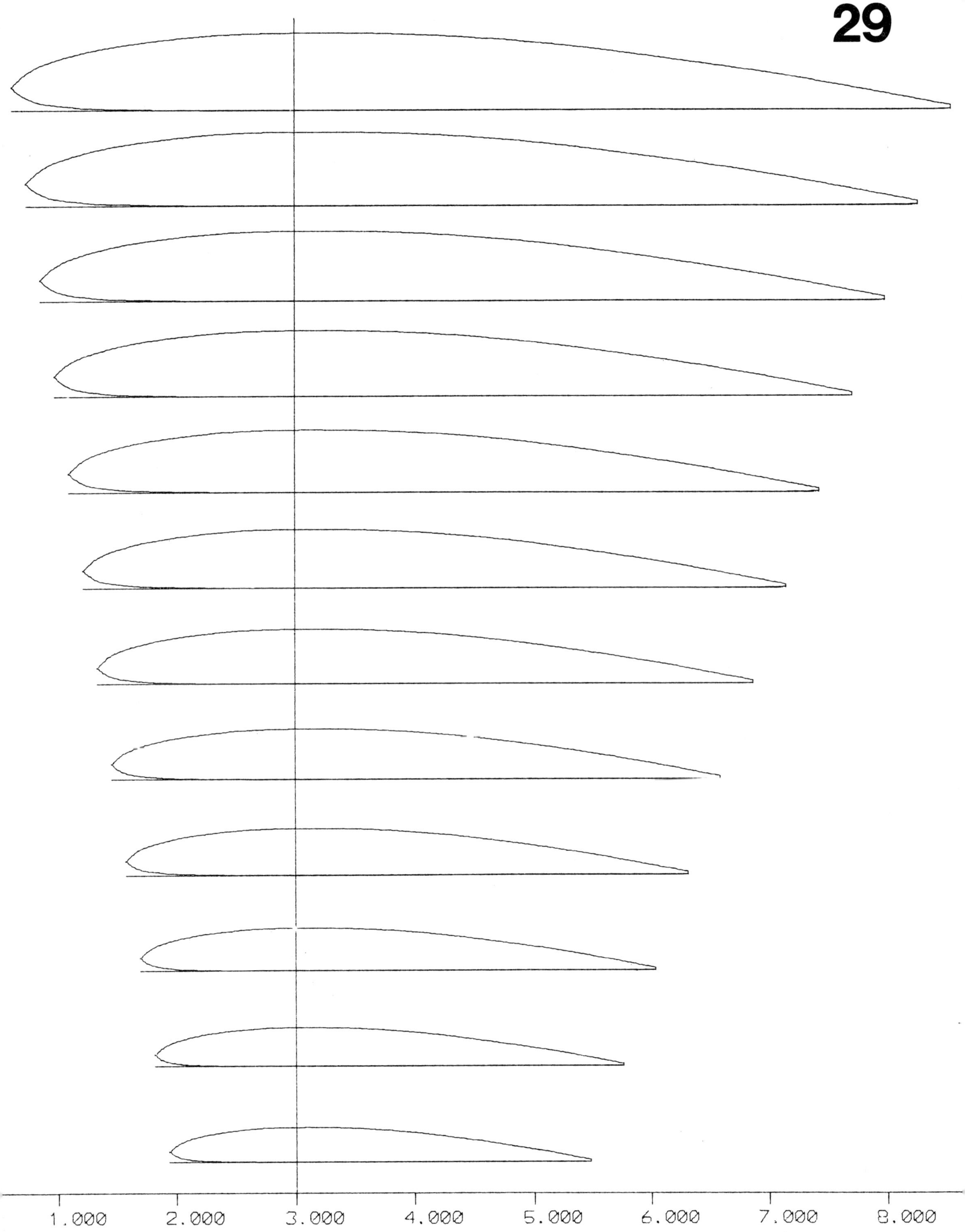

GOTTINGEN 795, SKIMMED

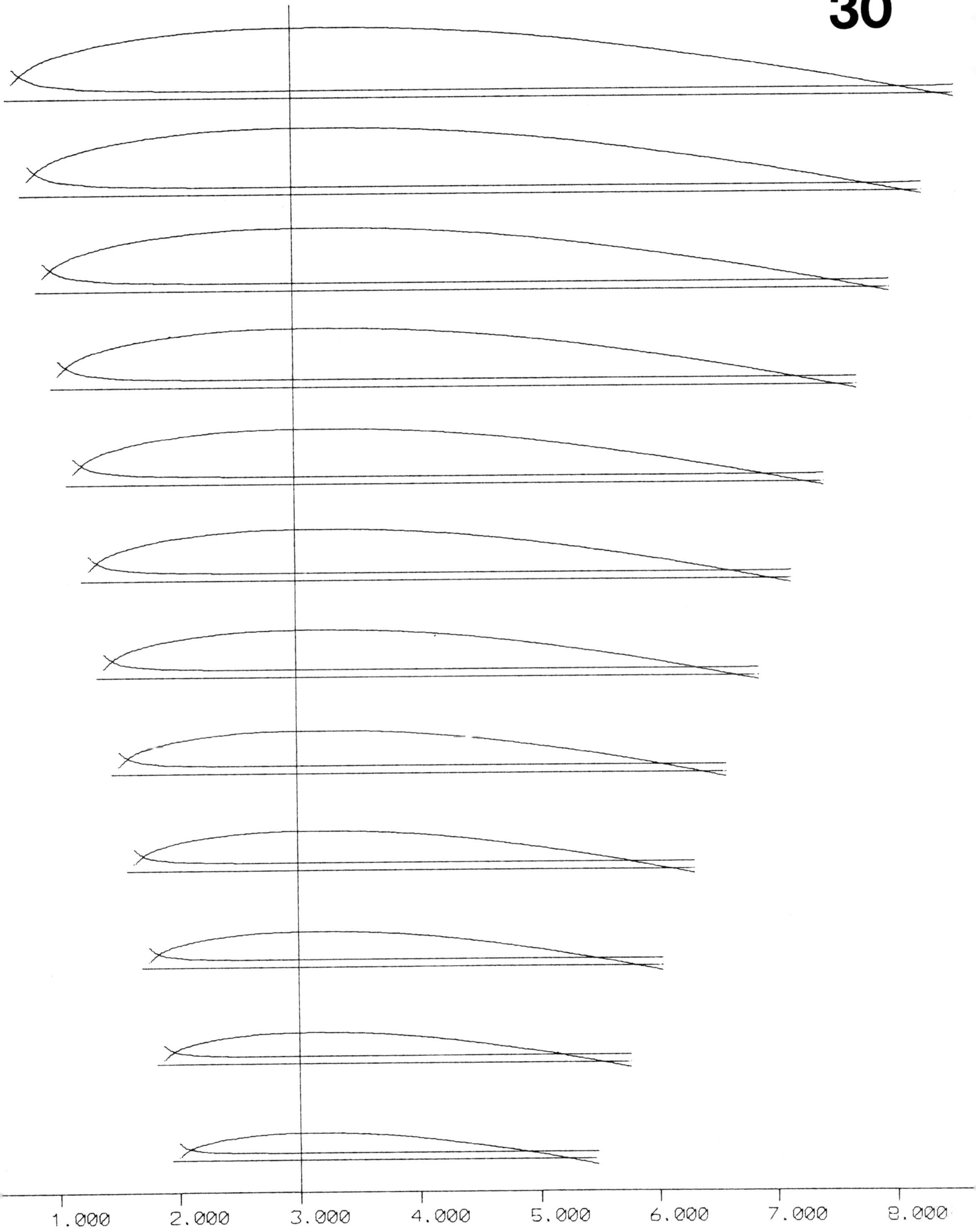

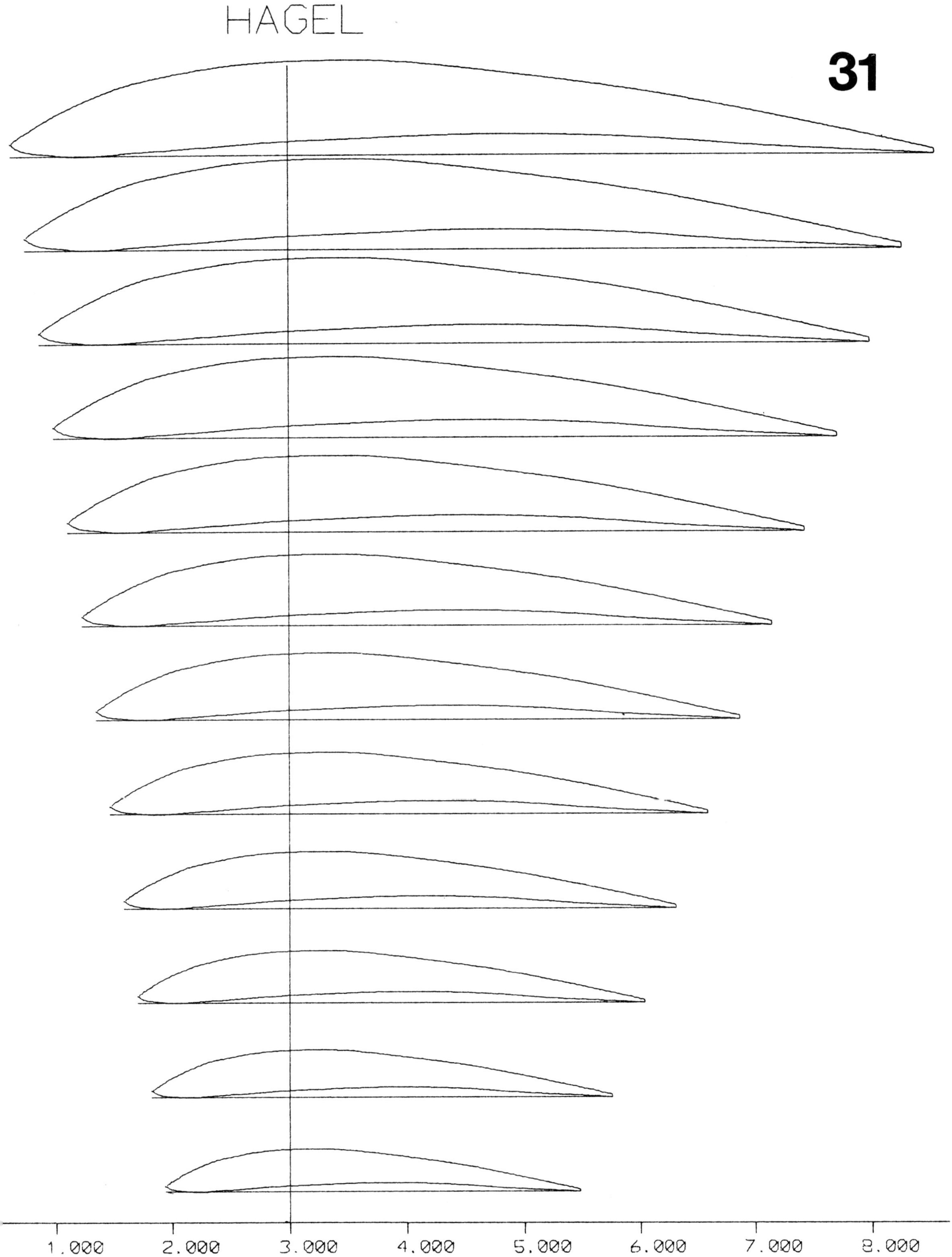
HAGEL
1.000
2.000
3.000
4.000
5.000
6.000
7.000
8.000

HAGEL, SKIMMED

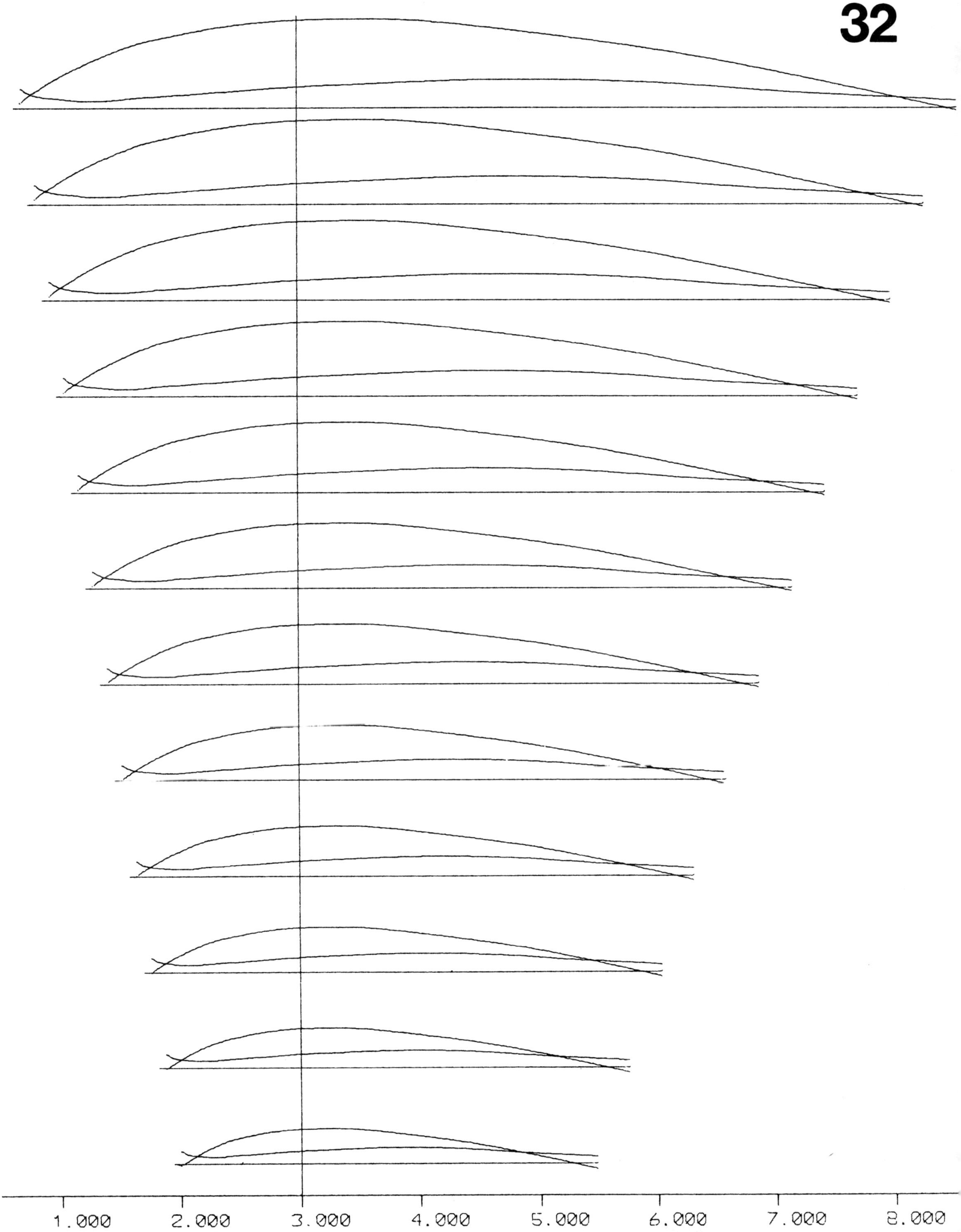

HAGEL BLENDED

33

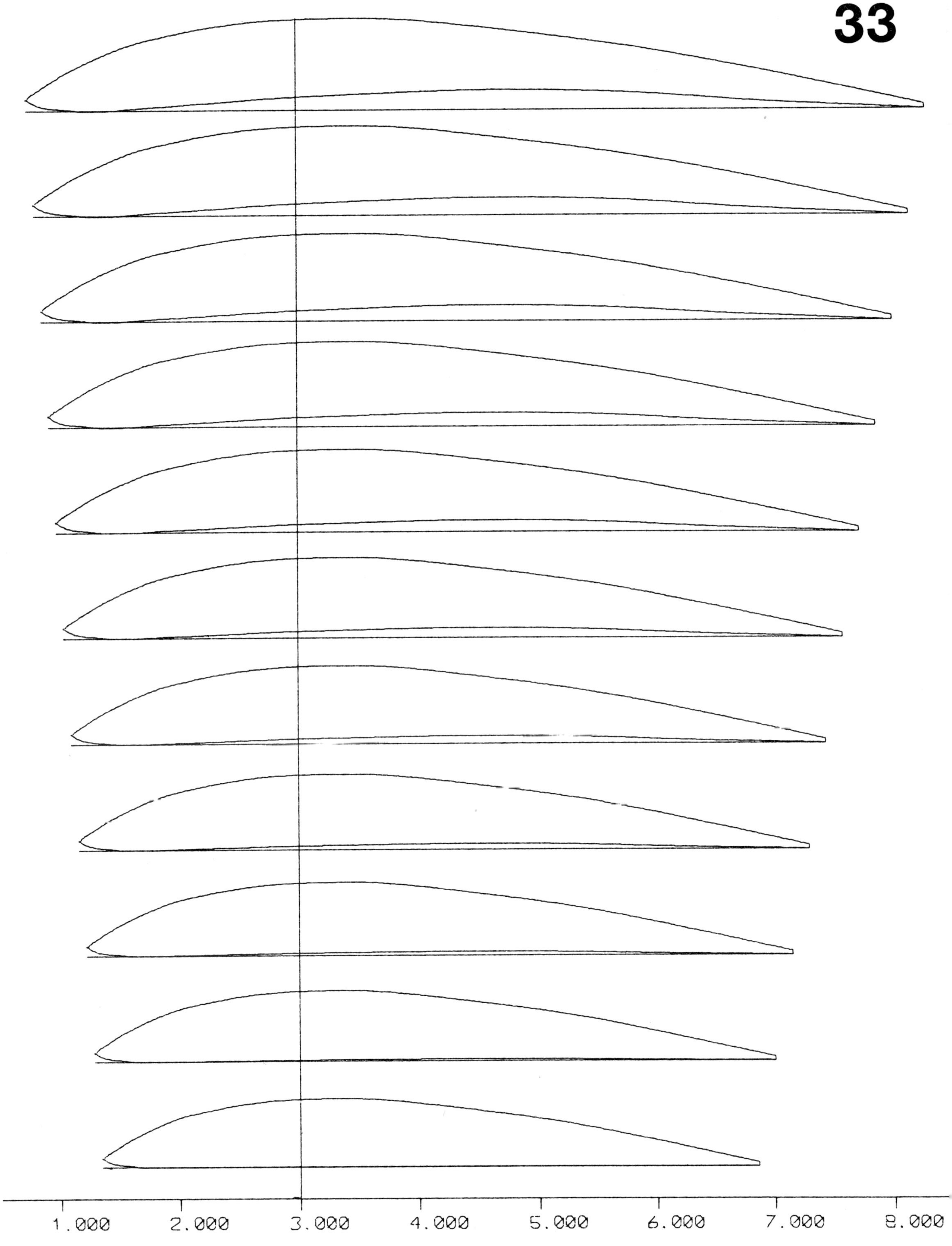

HAGEL BLENDED, SKIMMED

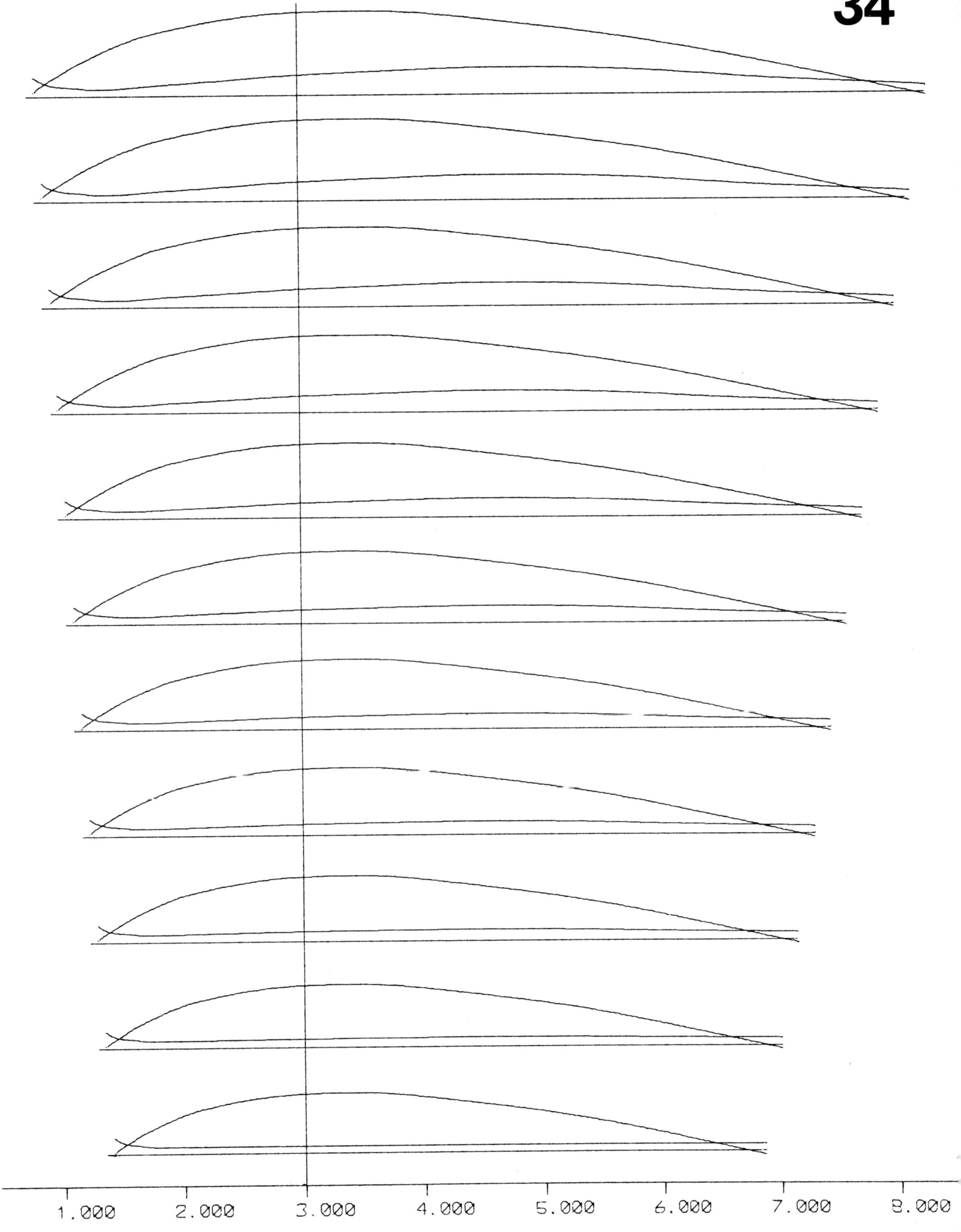

HAGEL BLENDED

35

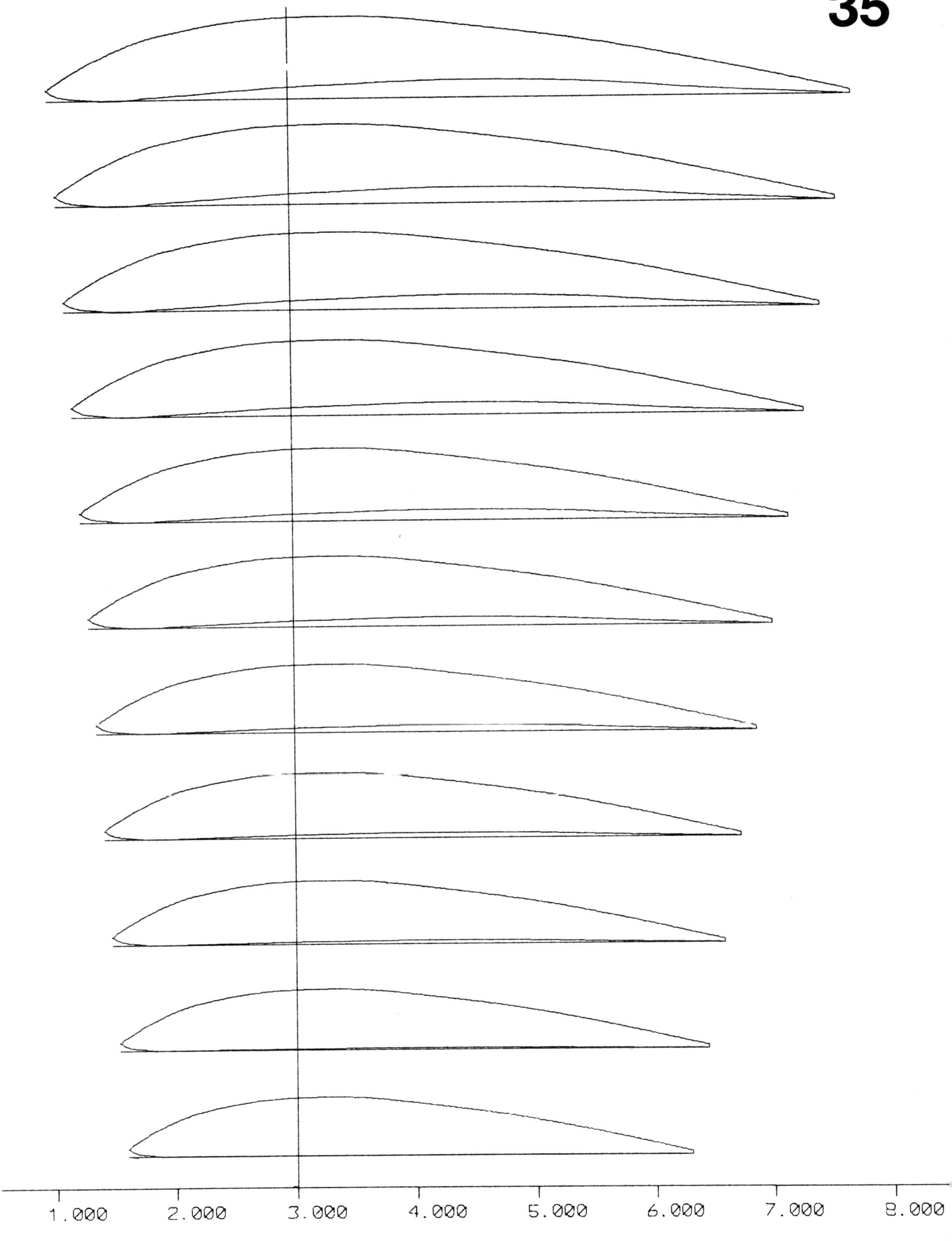

HAGEL BLENDED, SKIMMED

36

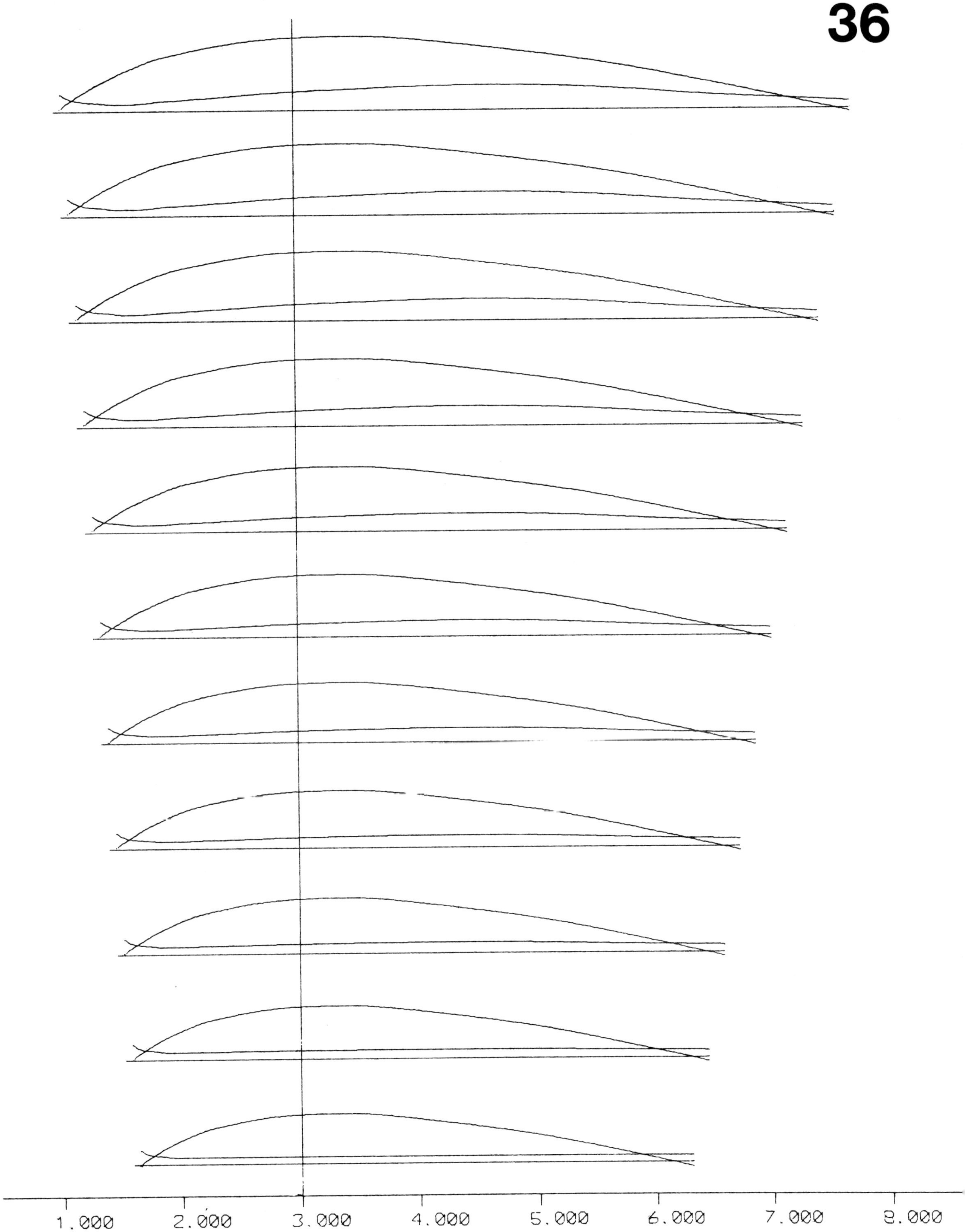

HANSEN AH-6-40-7

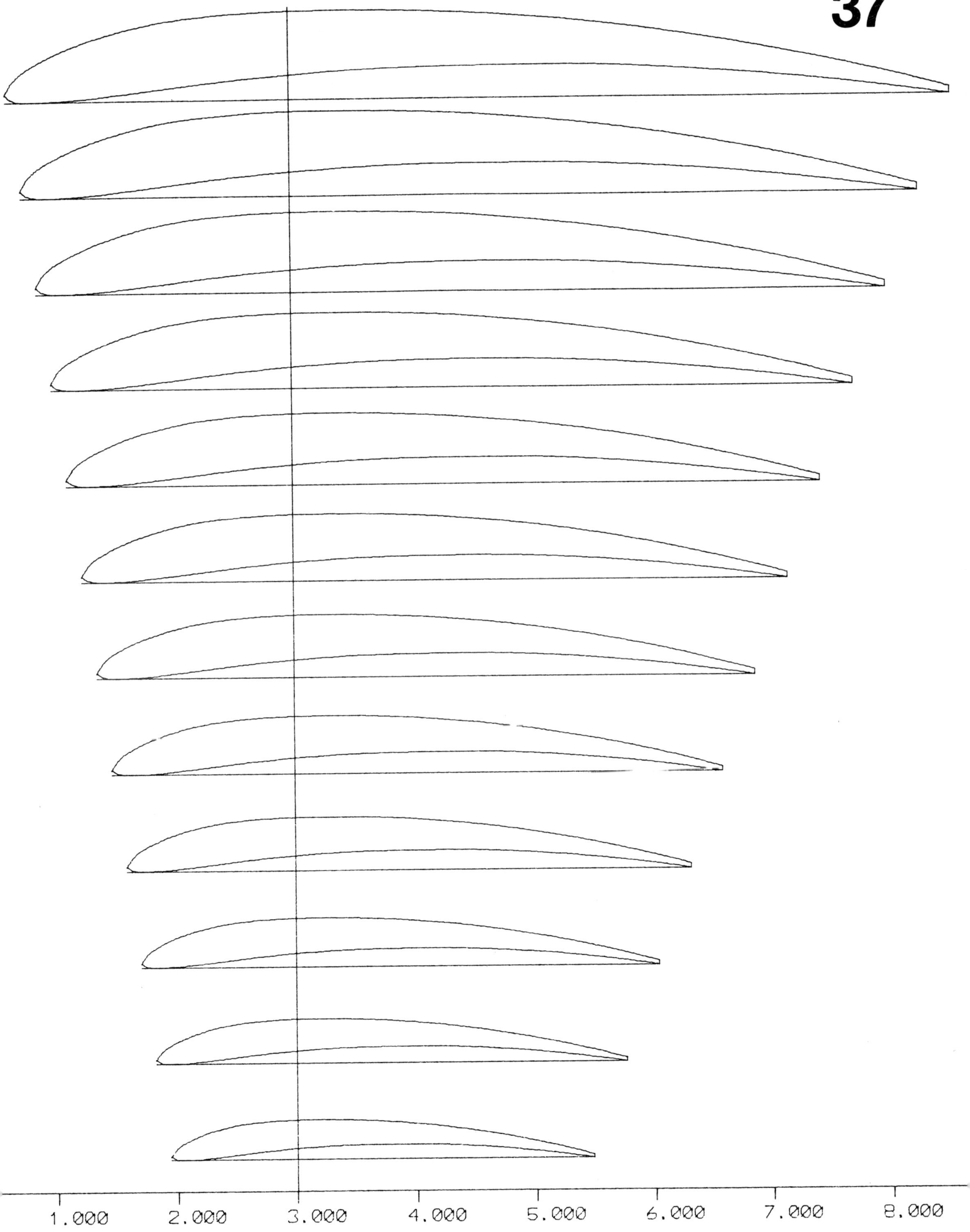

HANSEN AH-6-40-7, SKIMMED

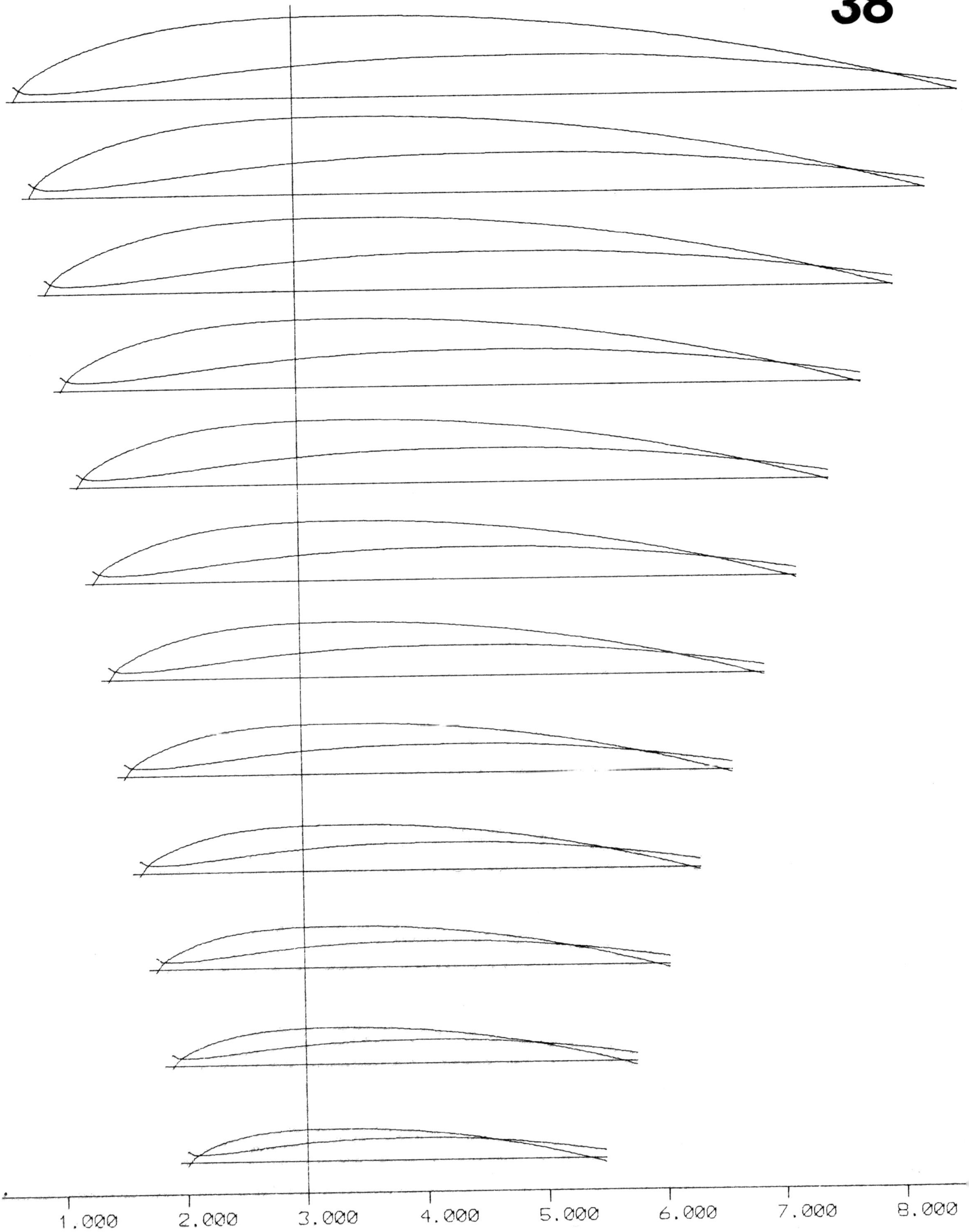

HANSEN AH-6-40-7 BLENDED

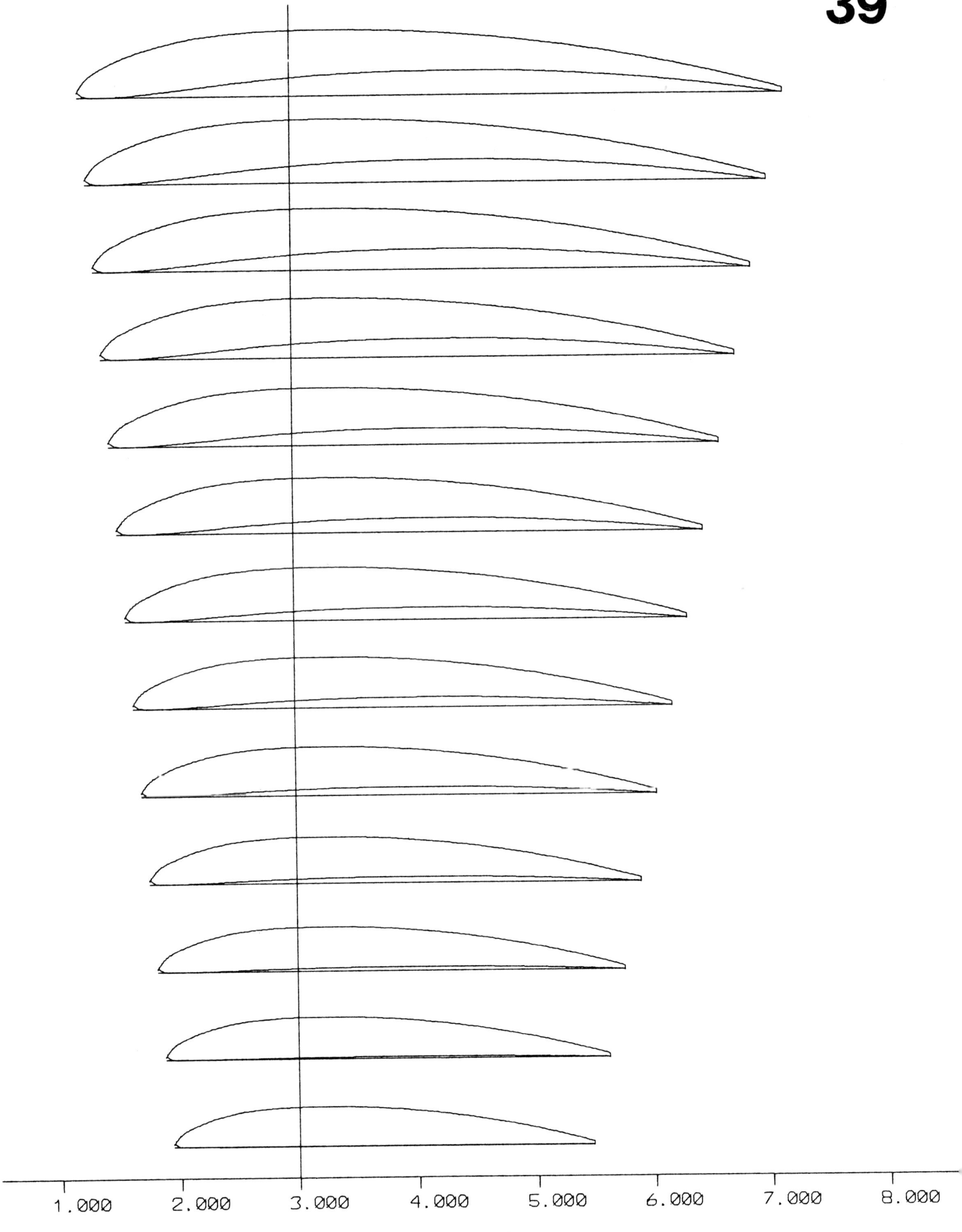

HANSEN AH-6-40-7 BLENDED, SKIMMED

40

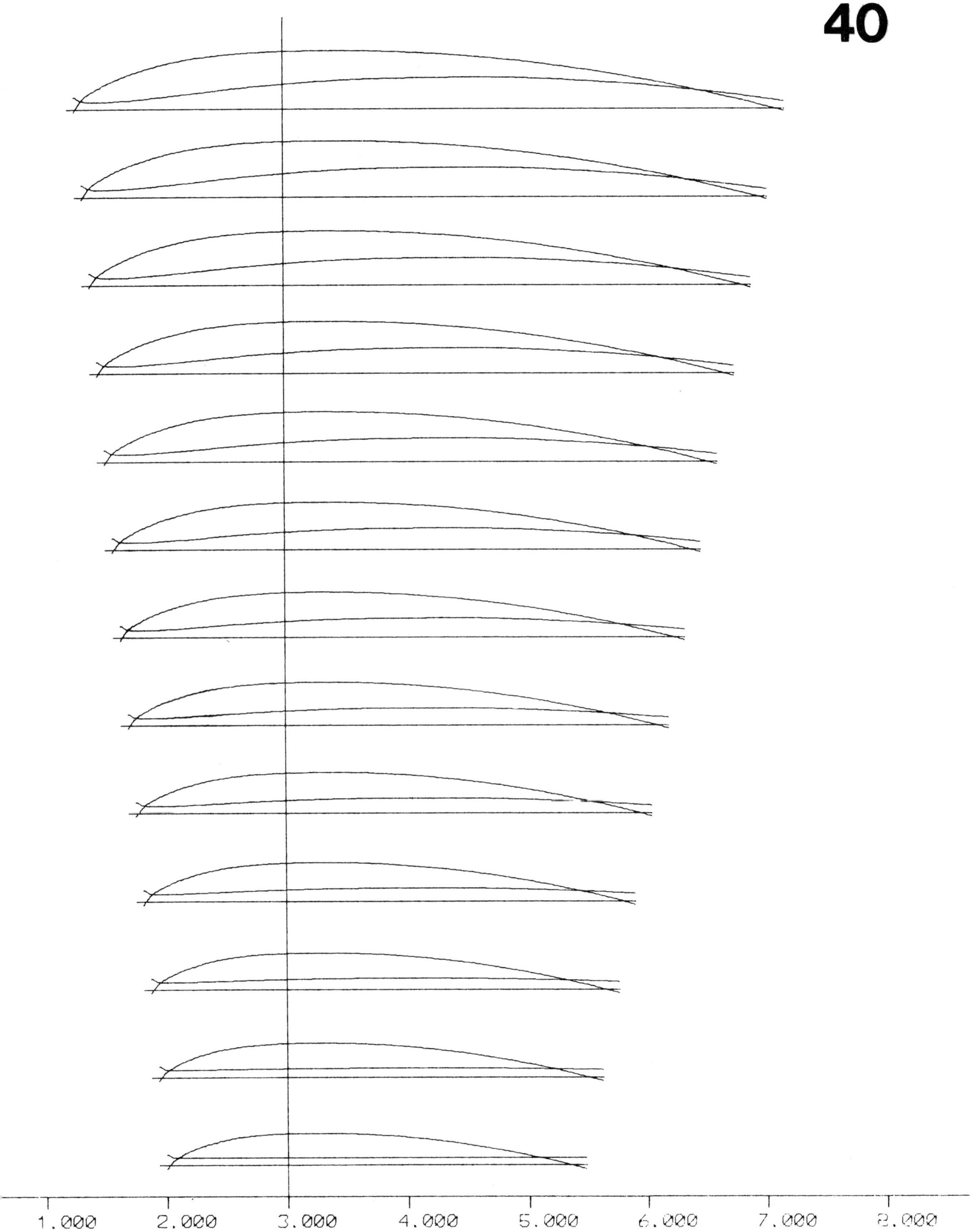

HANSEN AH-6-40-7 BLENDED

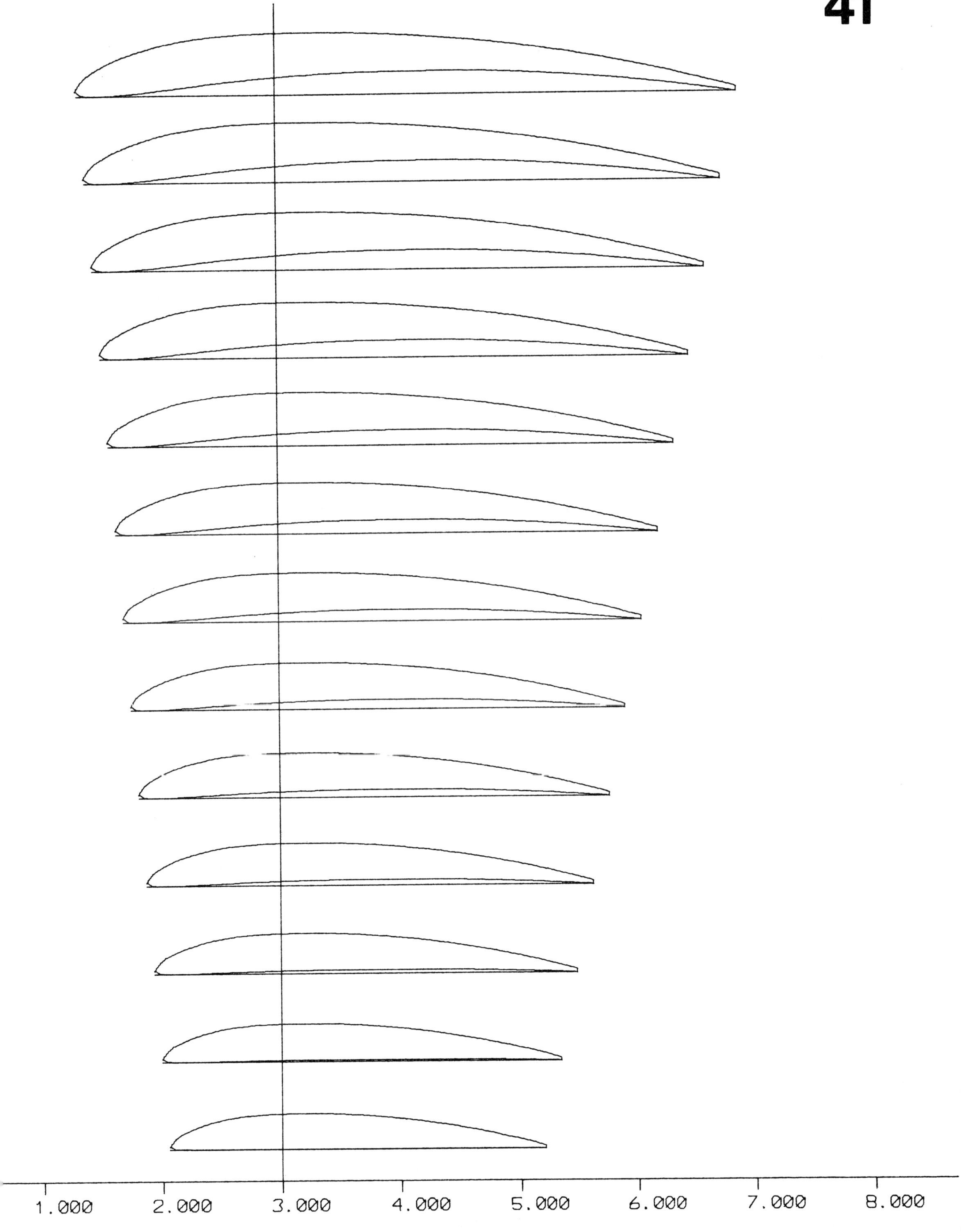

HANSEN AH-6-40-7 BLENDED, SKIMMED

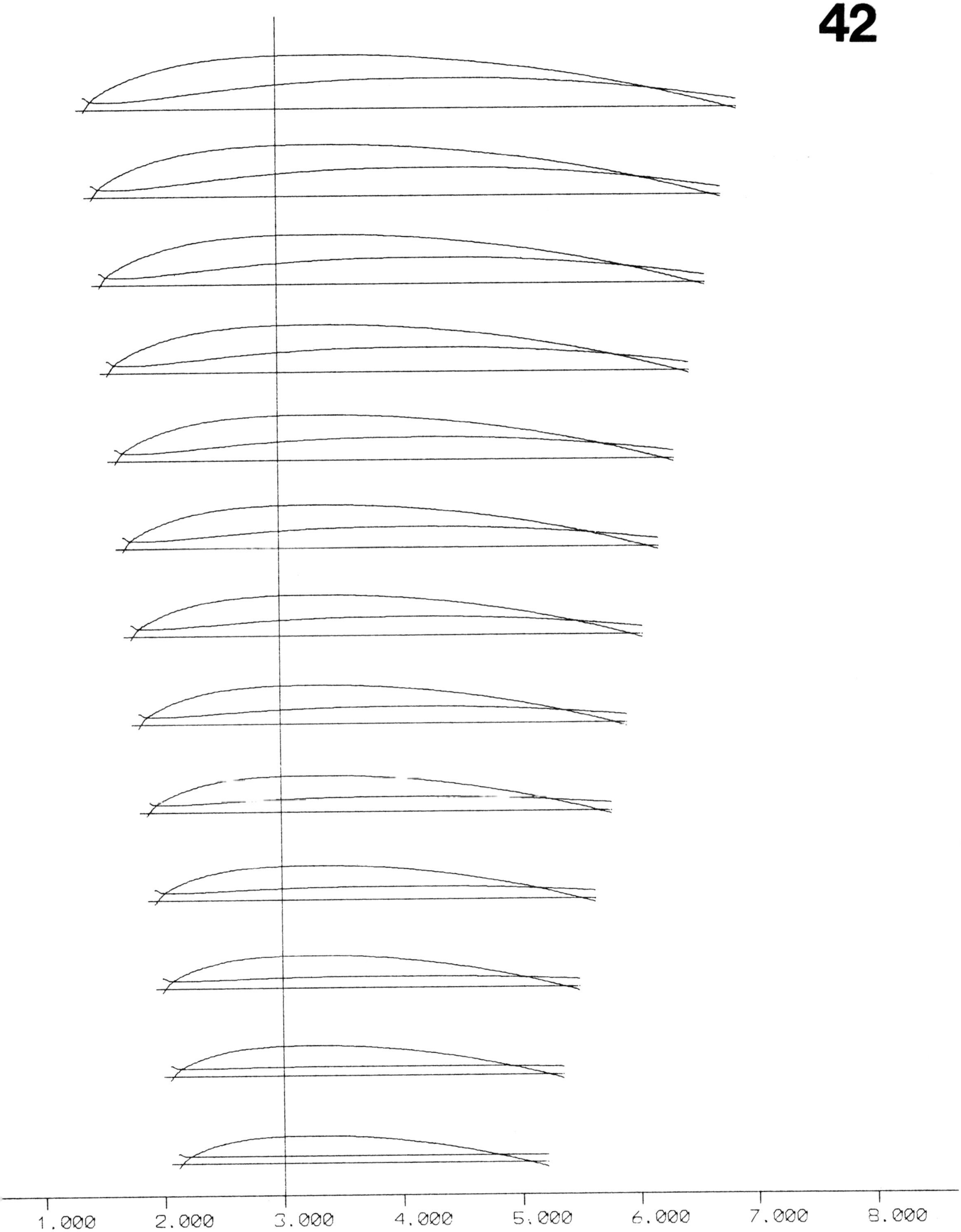

HANSEN AH-7-47-6

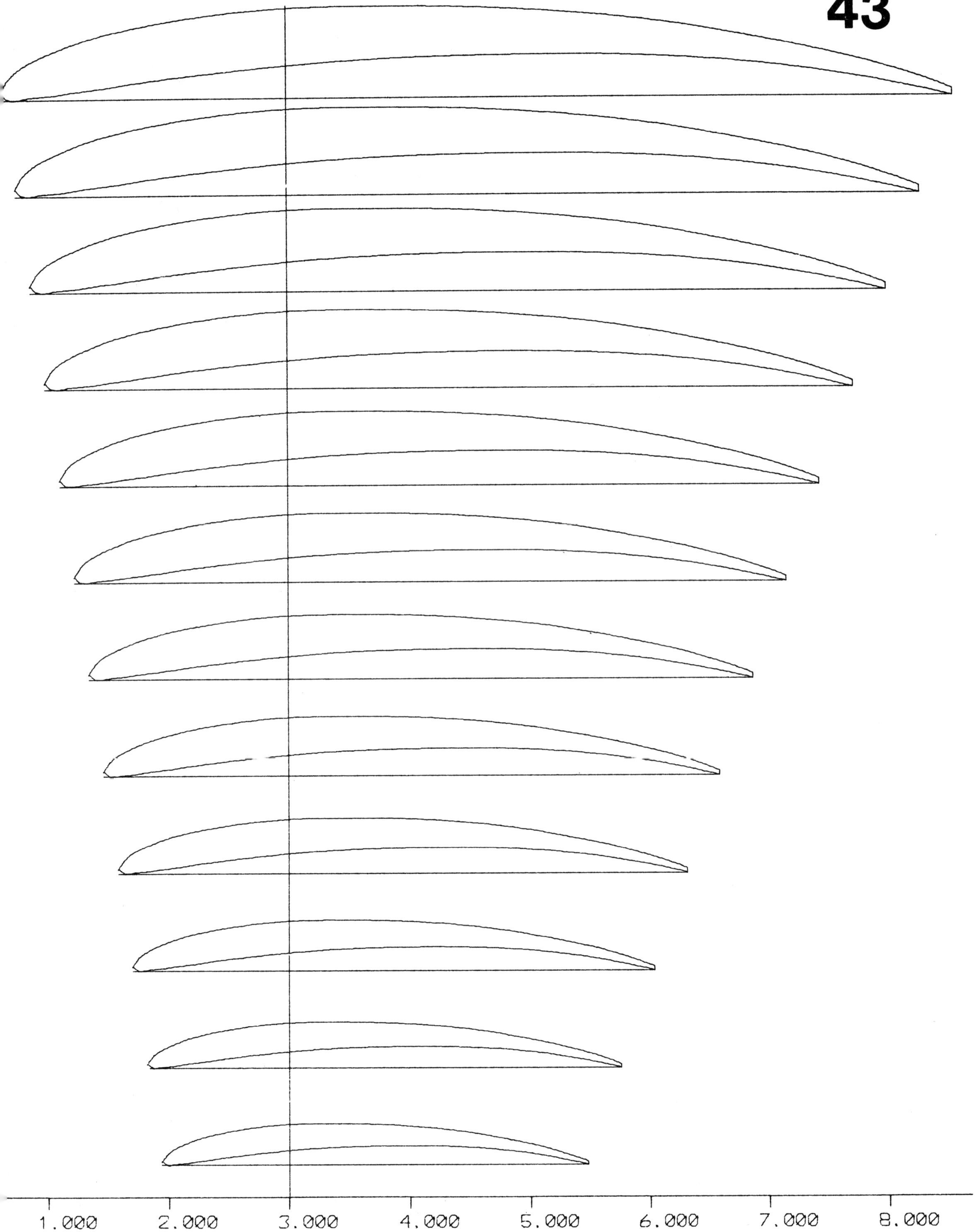

HANSEN AH-7-47-6, SKIMMED

44

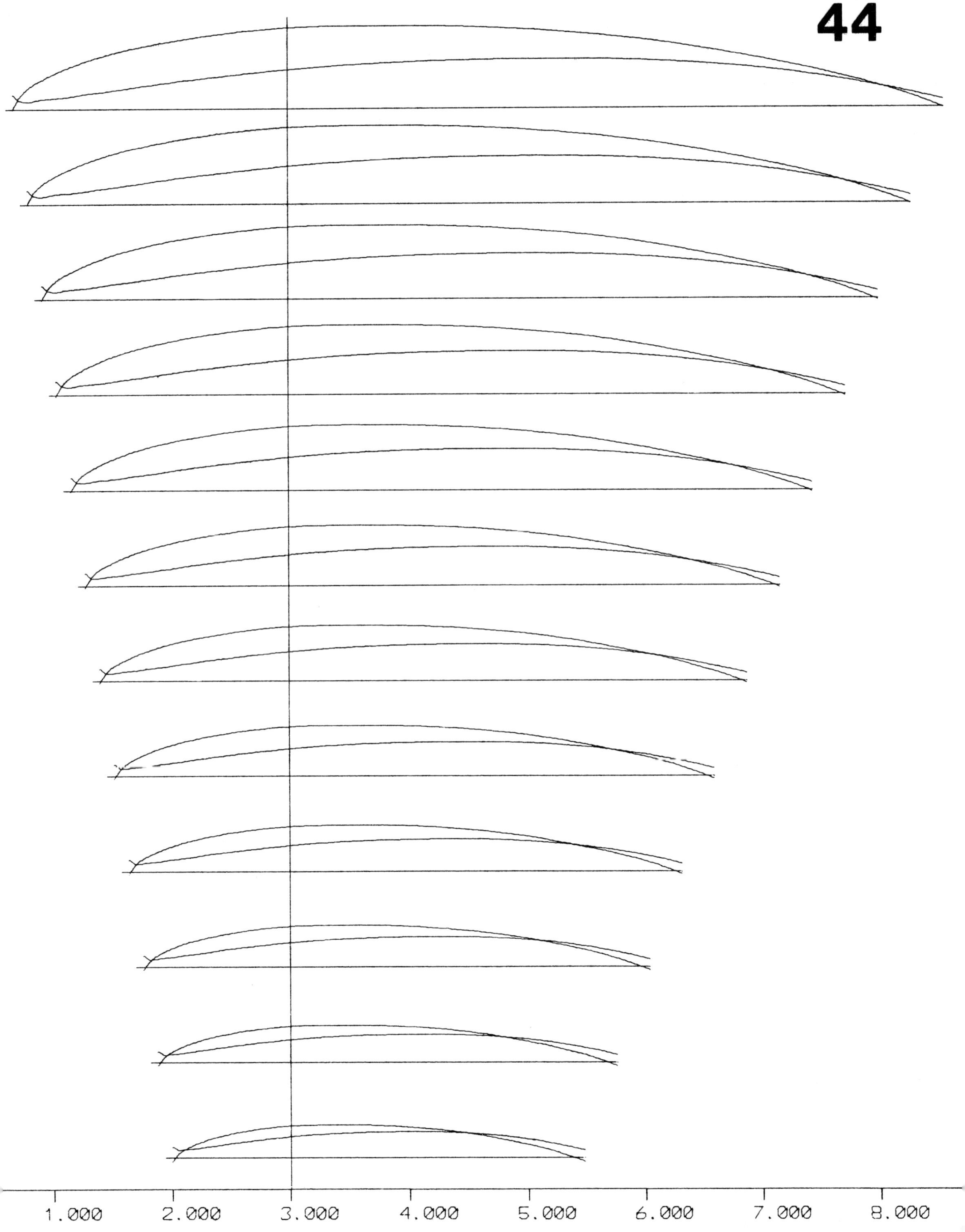

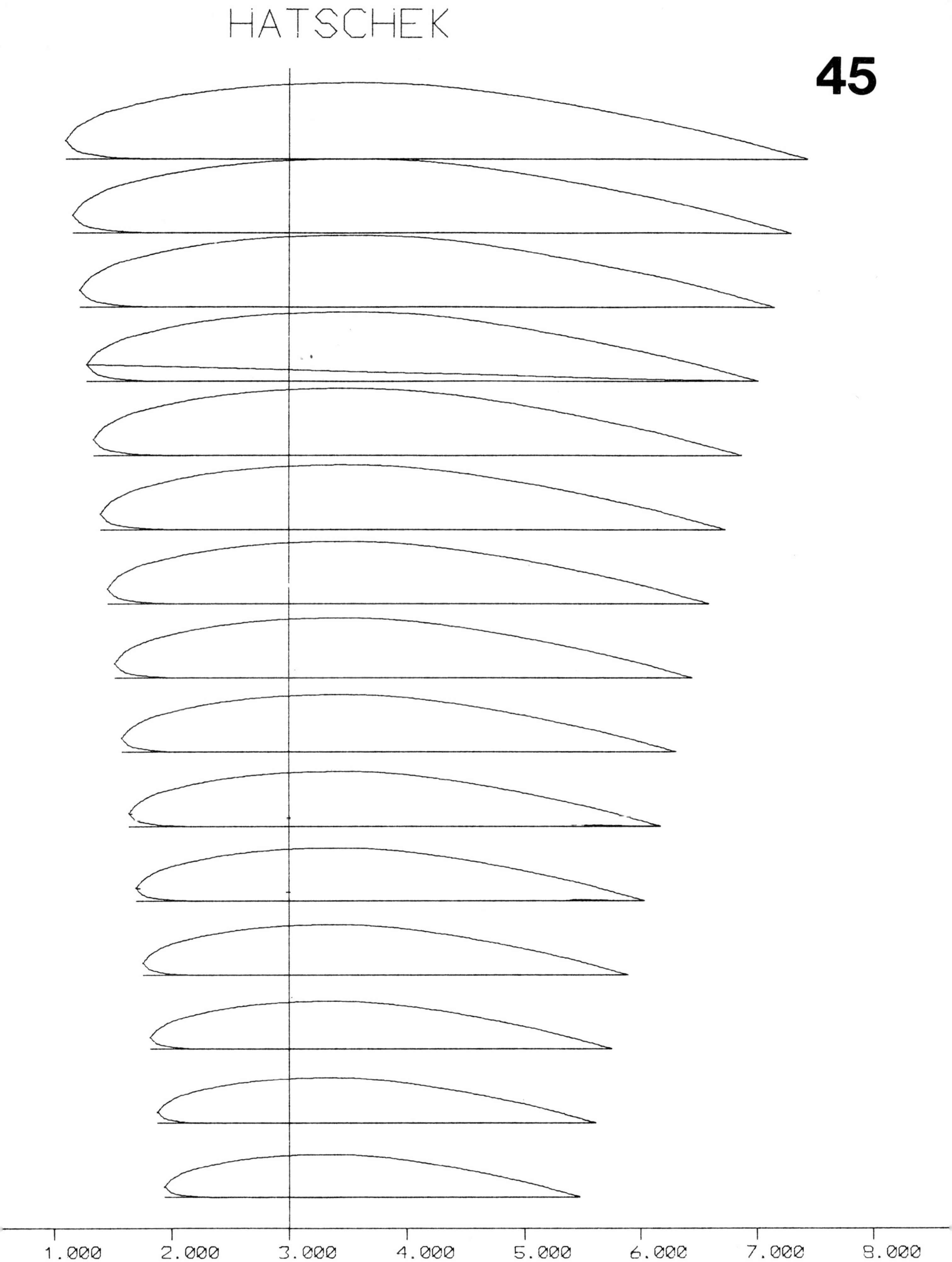
HATSCHEK
1.000
2.000
3.000
4.000
5.000
6.000
7.000
8.000

HOREJSI

46

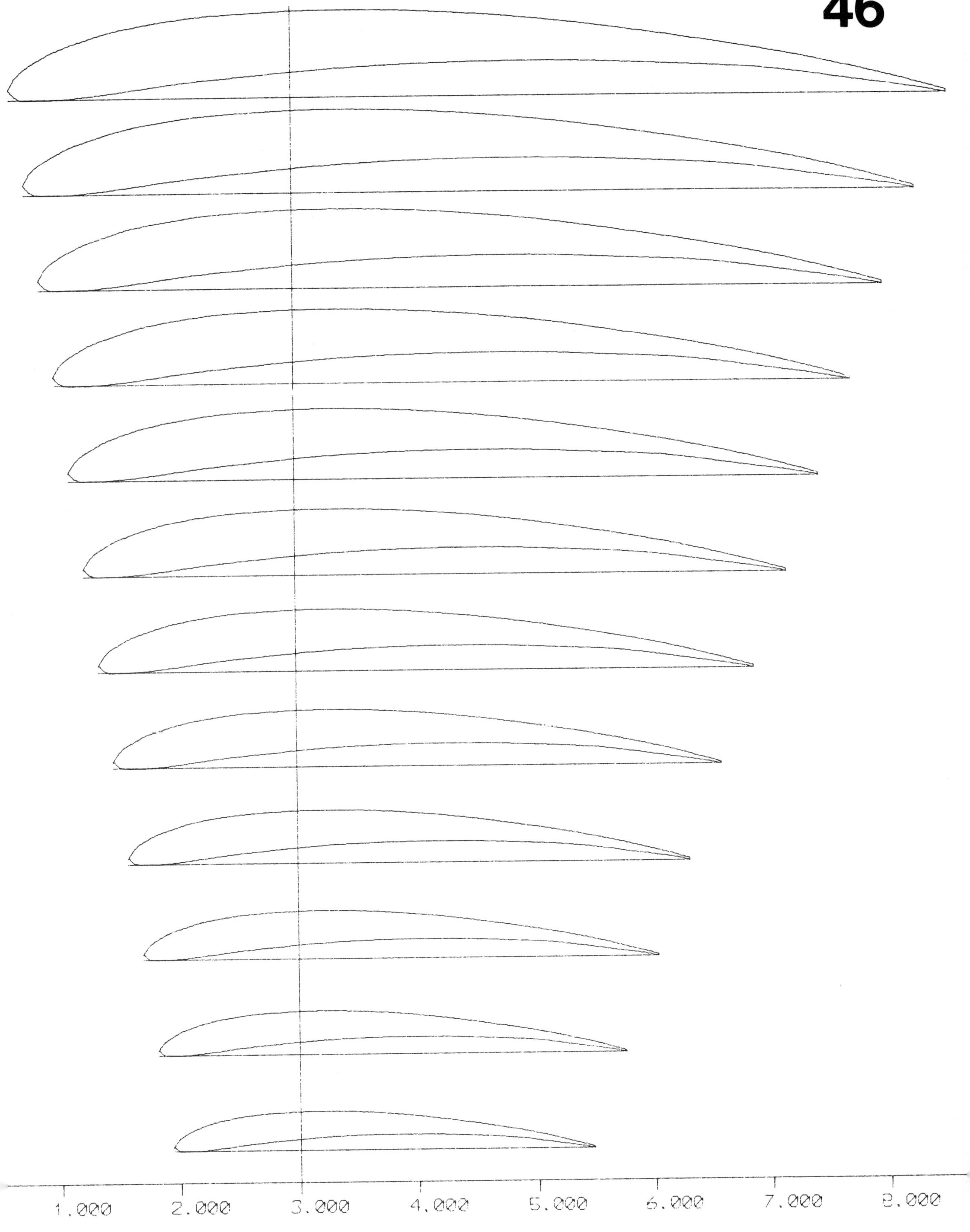

ISAACSON 64009

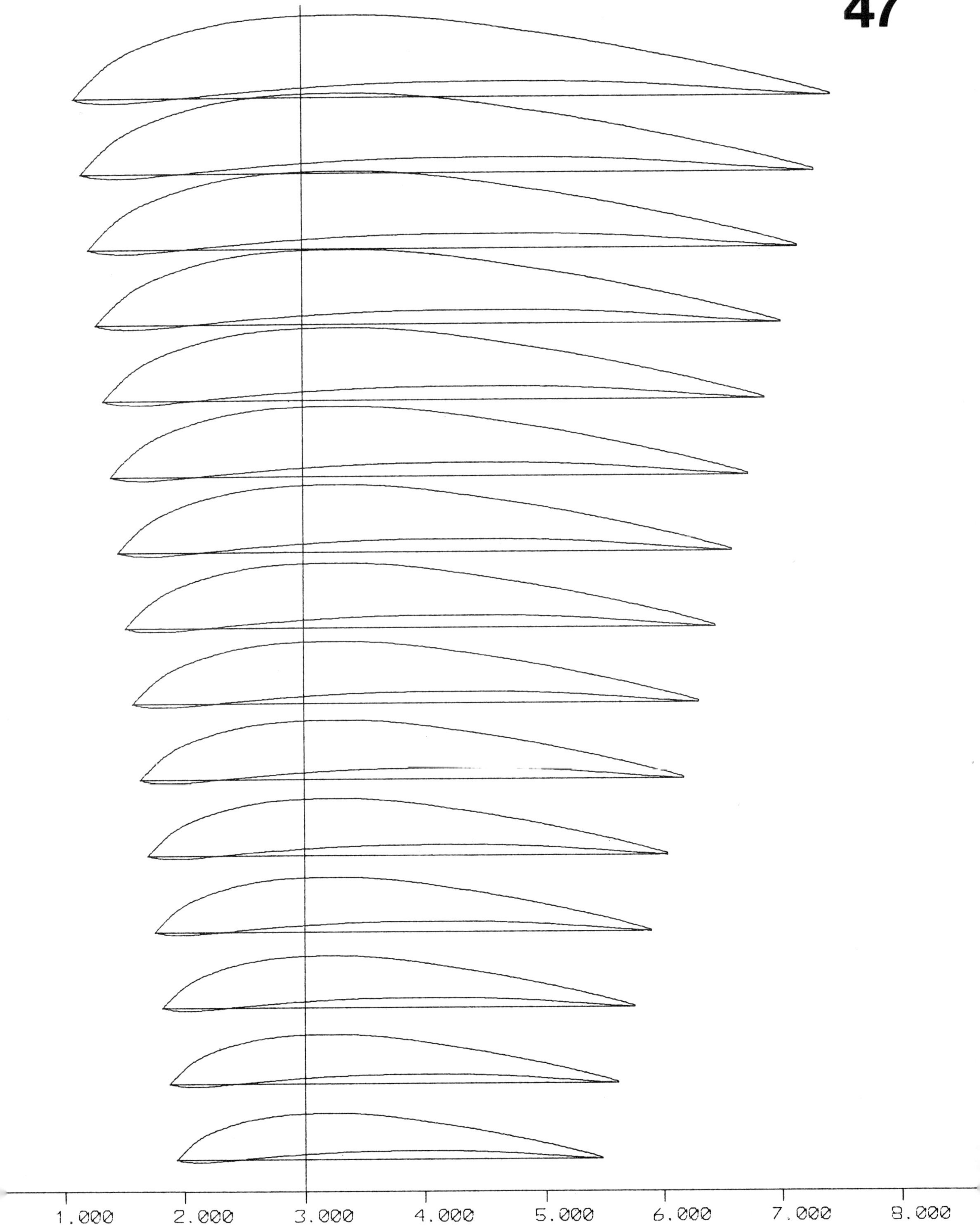

KACZANOWSKI G F 6

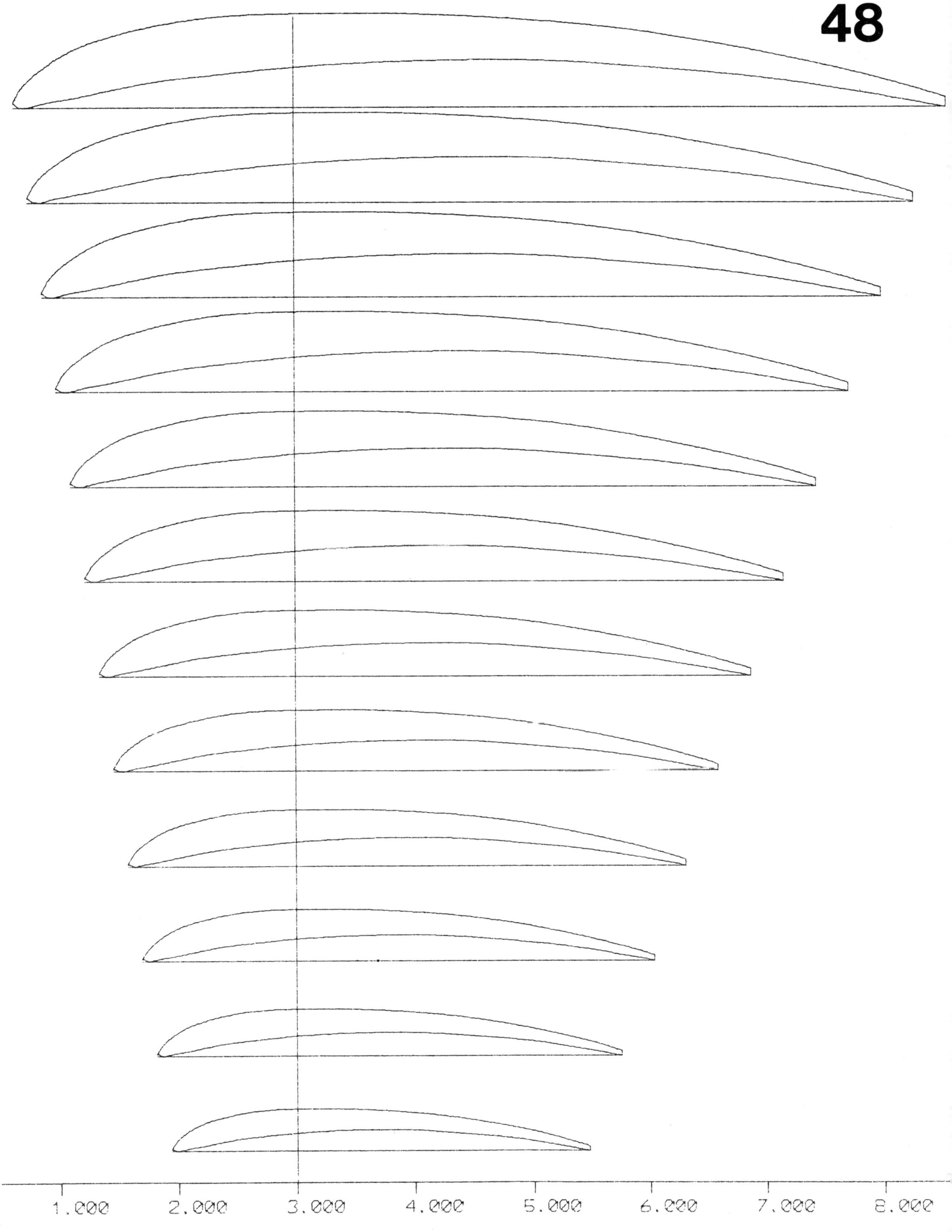

KACZANOWSKI G.F.6, SKIMMED

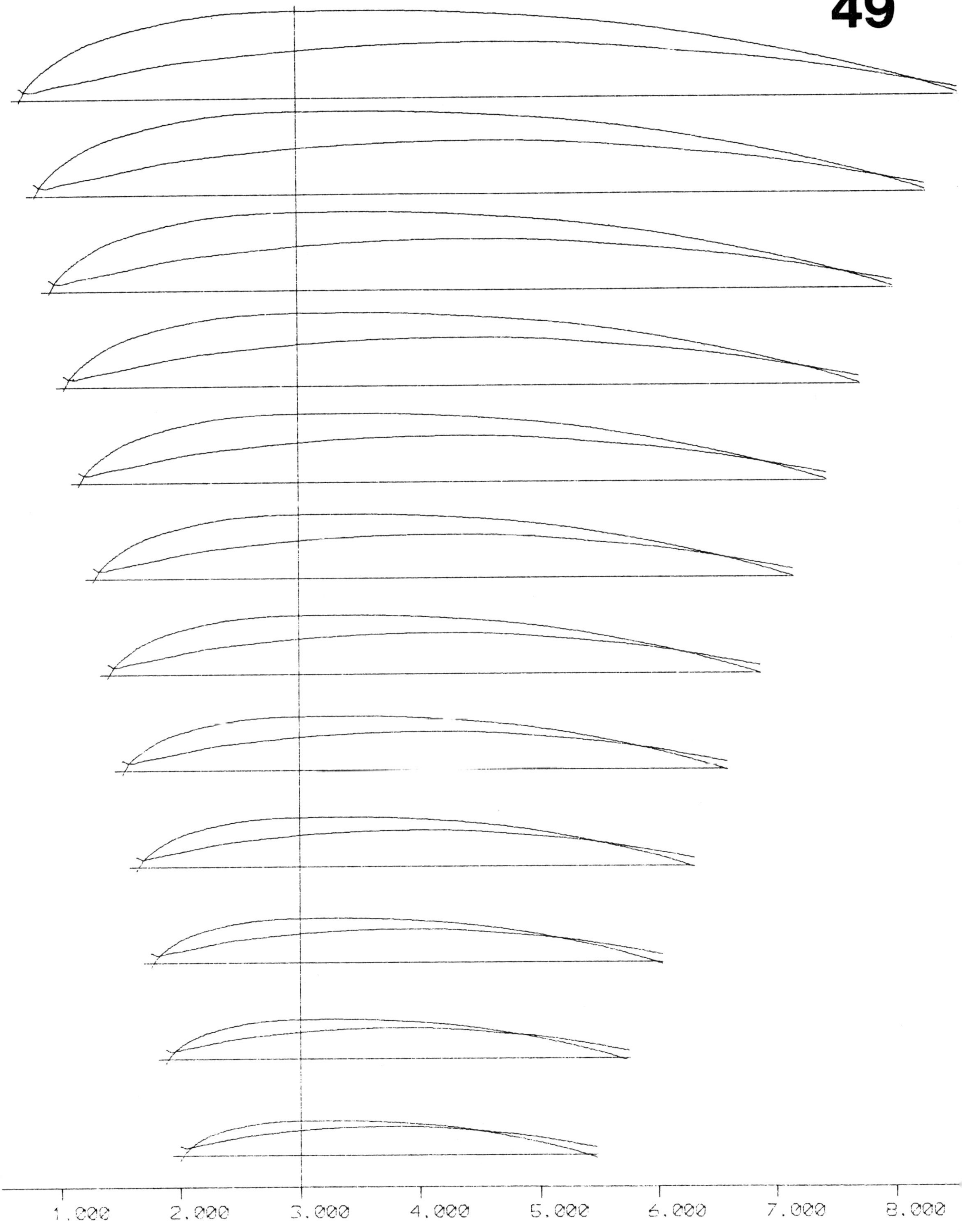

KOSTER 66

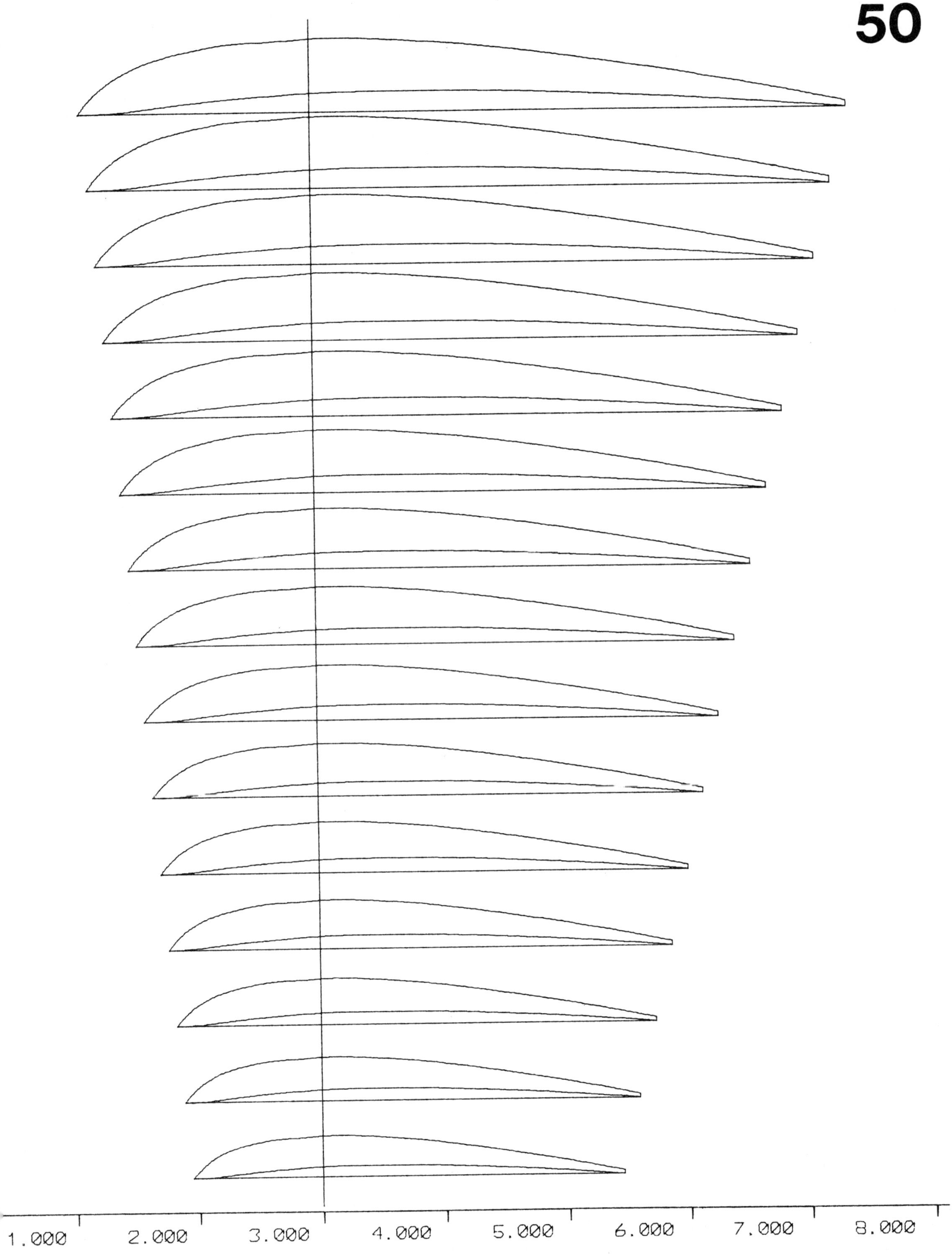

KOSTER 66, SKIMMED

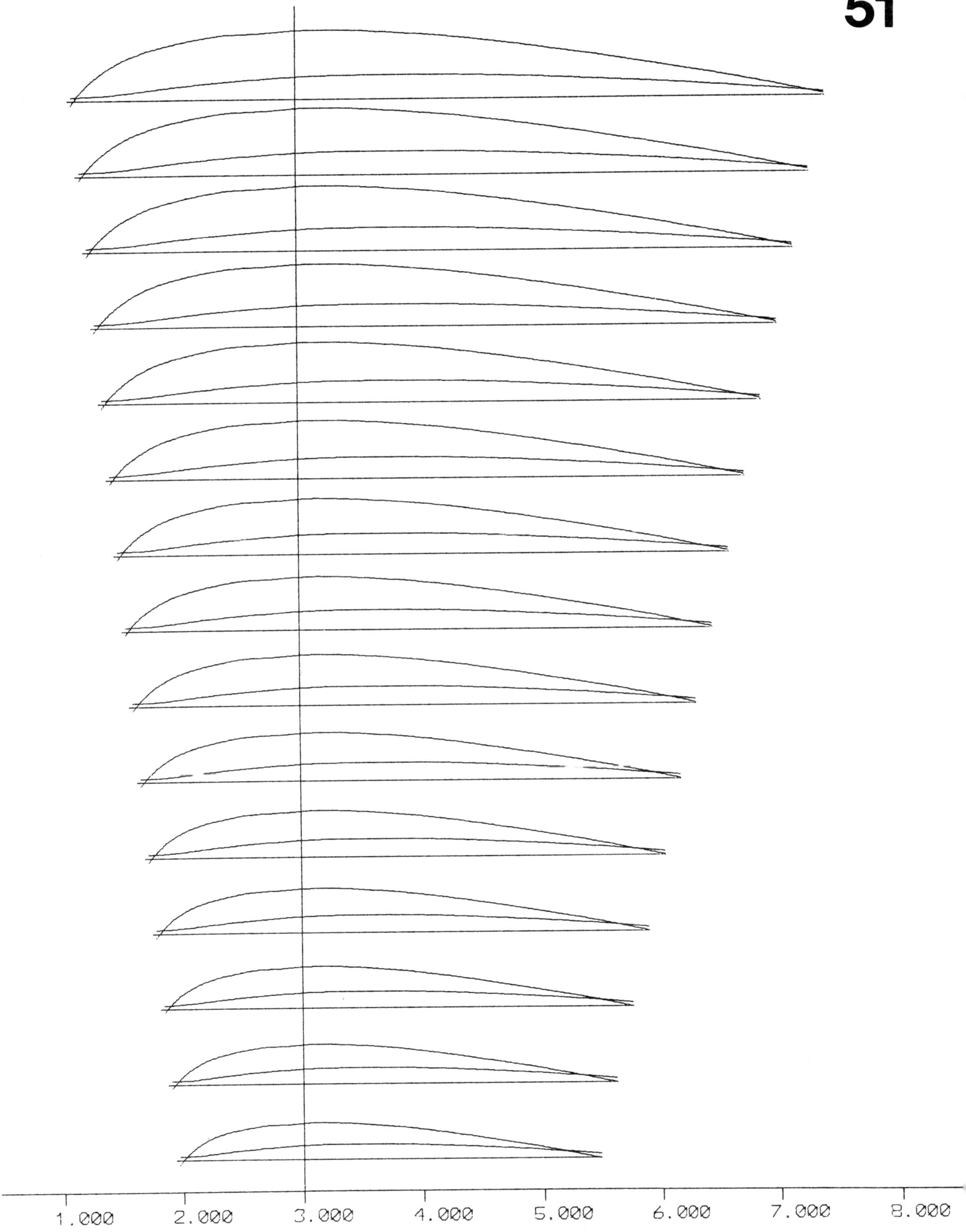

KOSTER 66 BLENDED

52

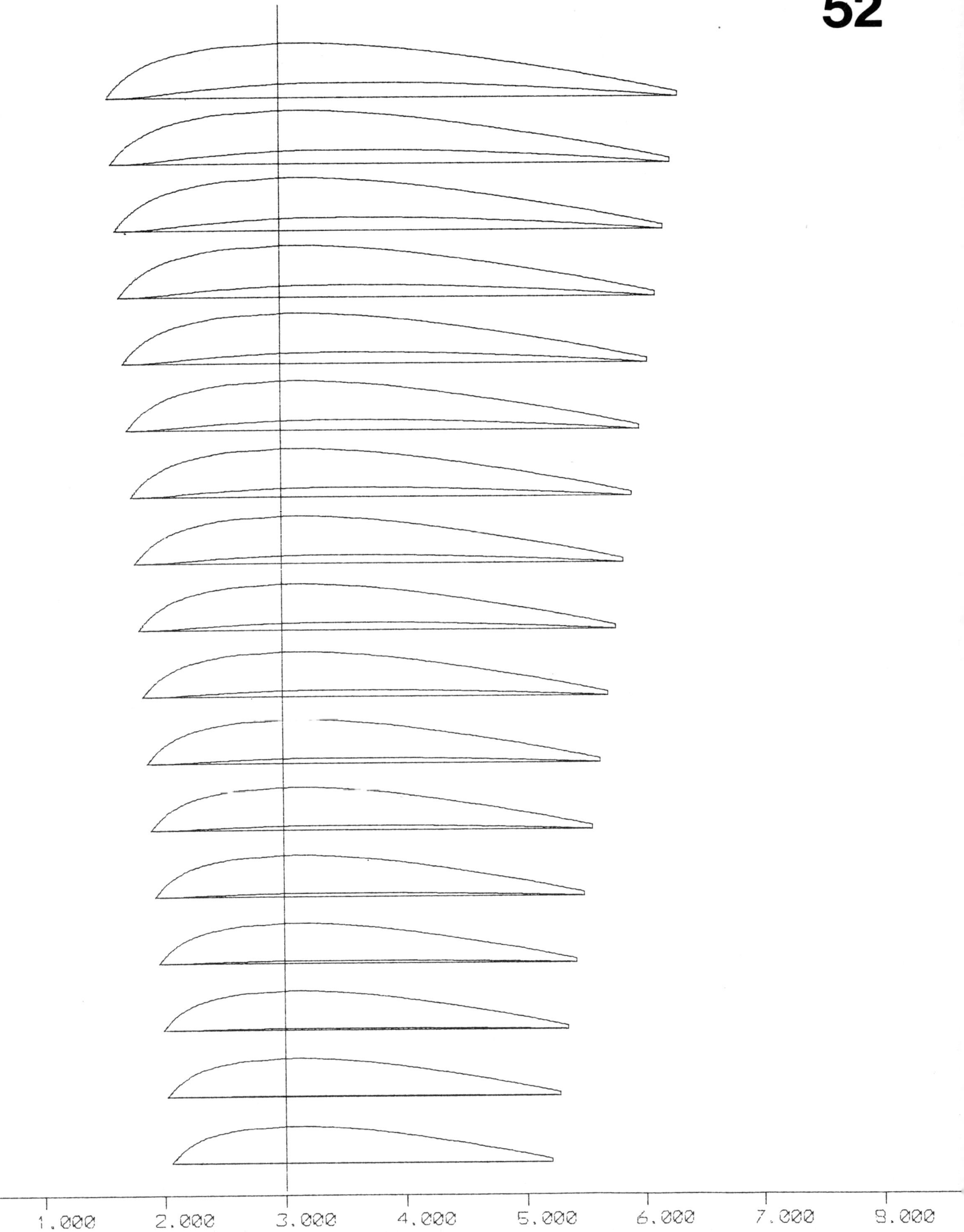

KOSTER 66 BLENDED, SKIMMED

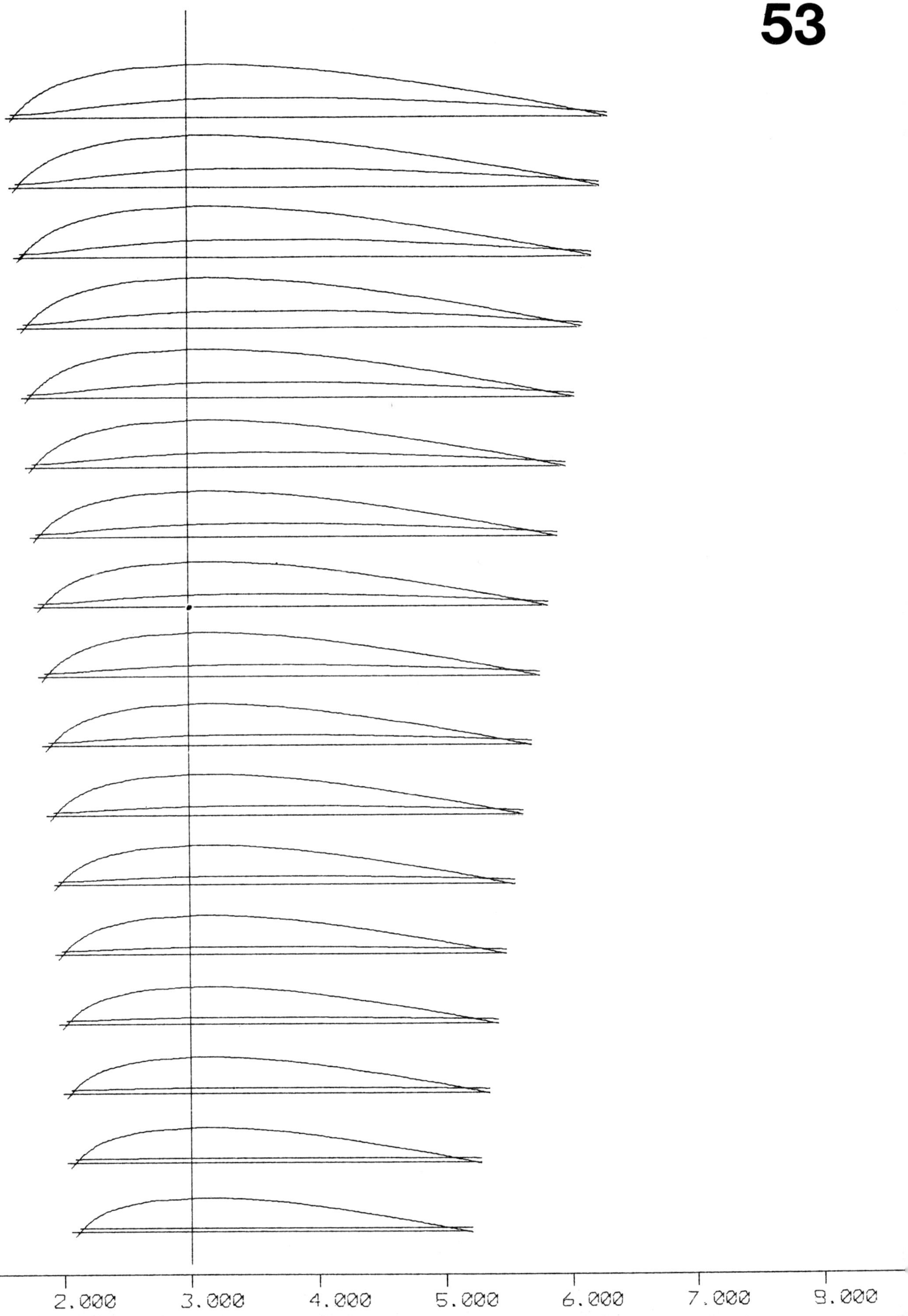

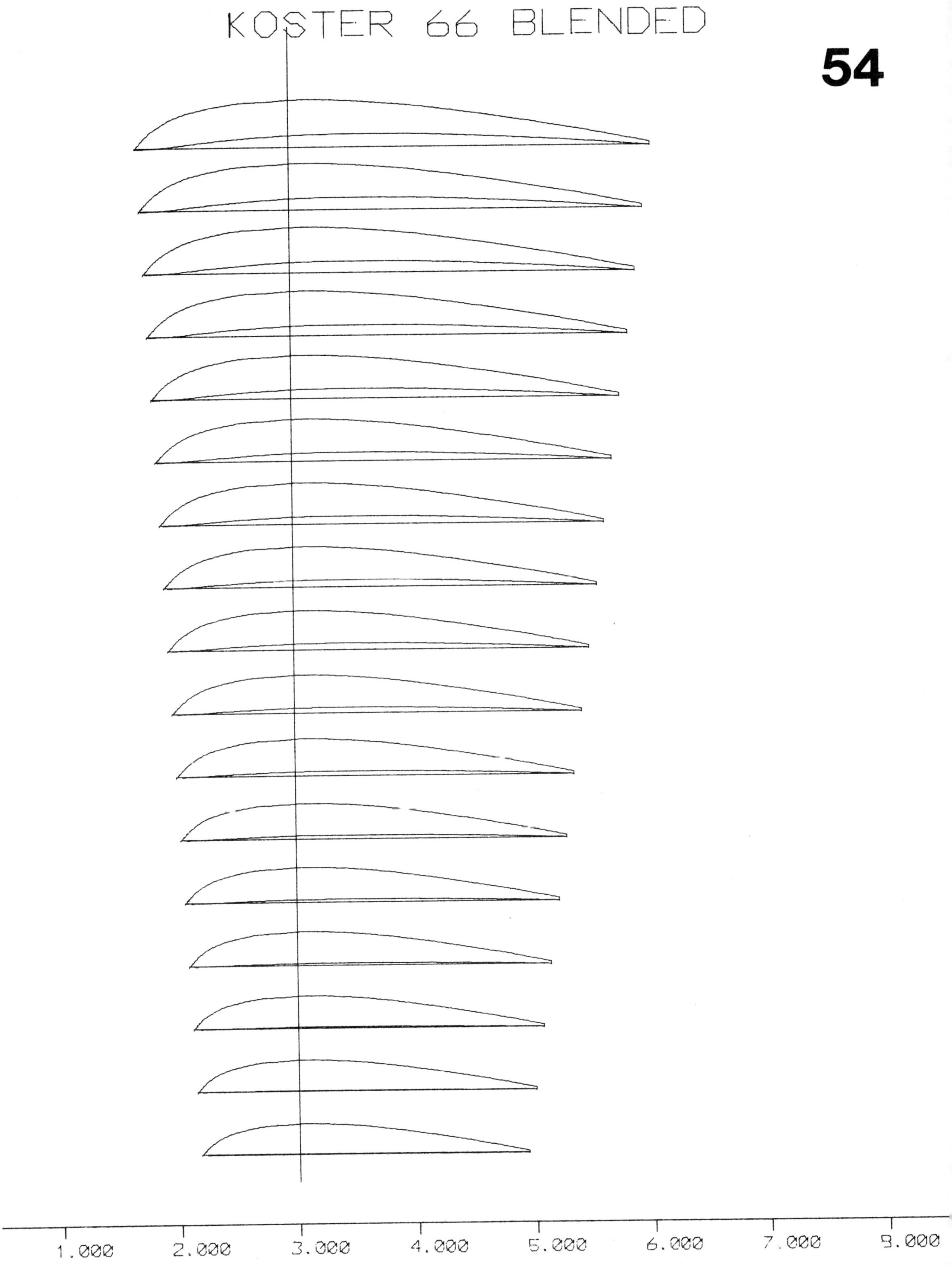
KOSTER 66 BLENDED
54
1.000
2.000
3.000
4.000
5.000
6.000
7.000
8.000

KOSTER 66 BLENDED, SKIMMED

55

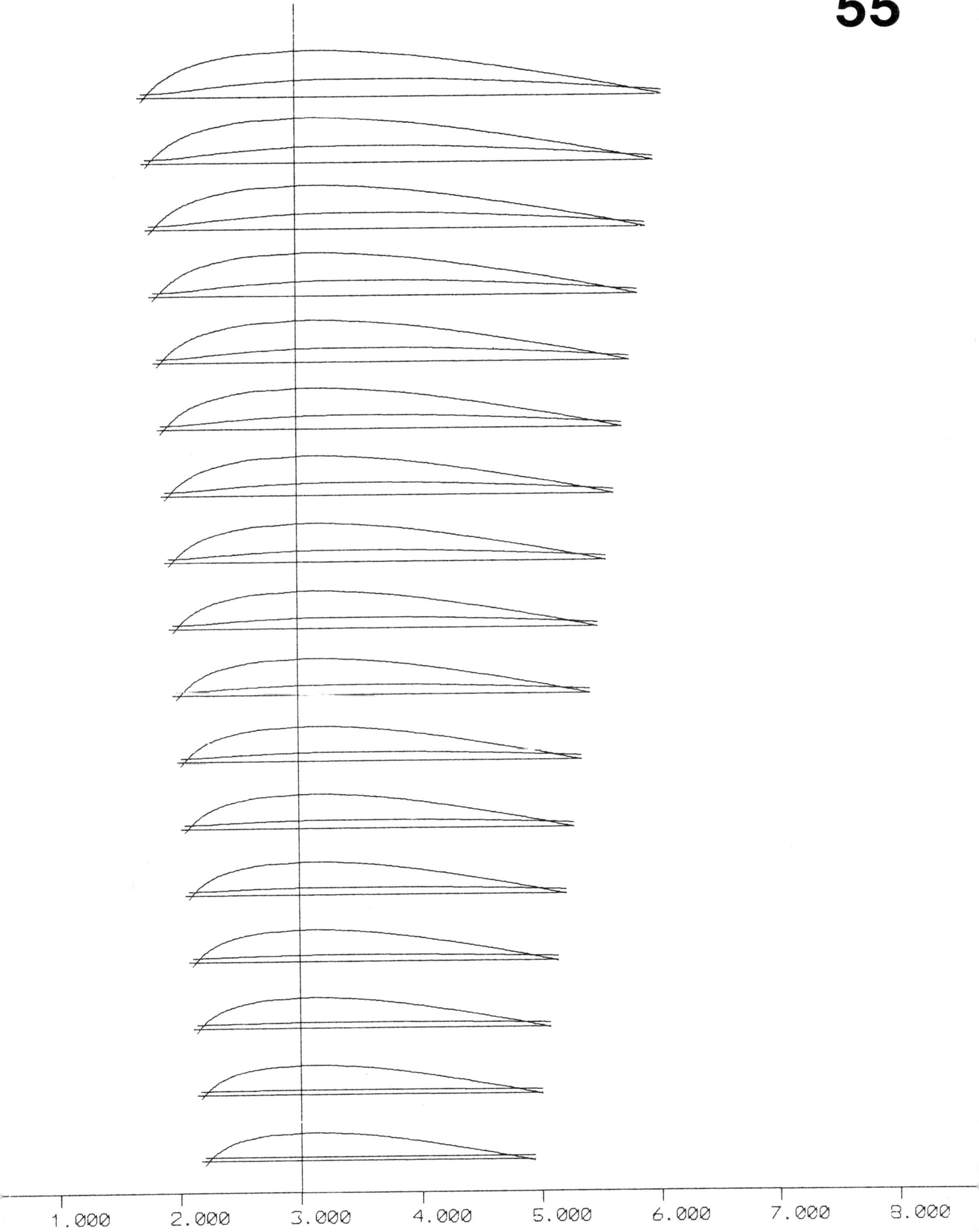

LINDNER

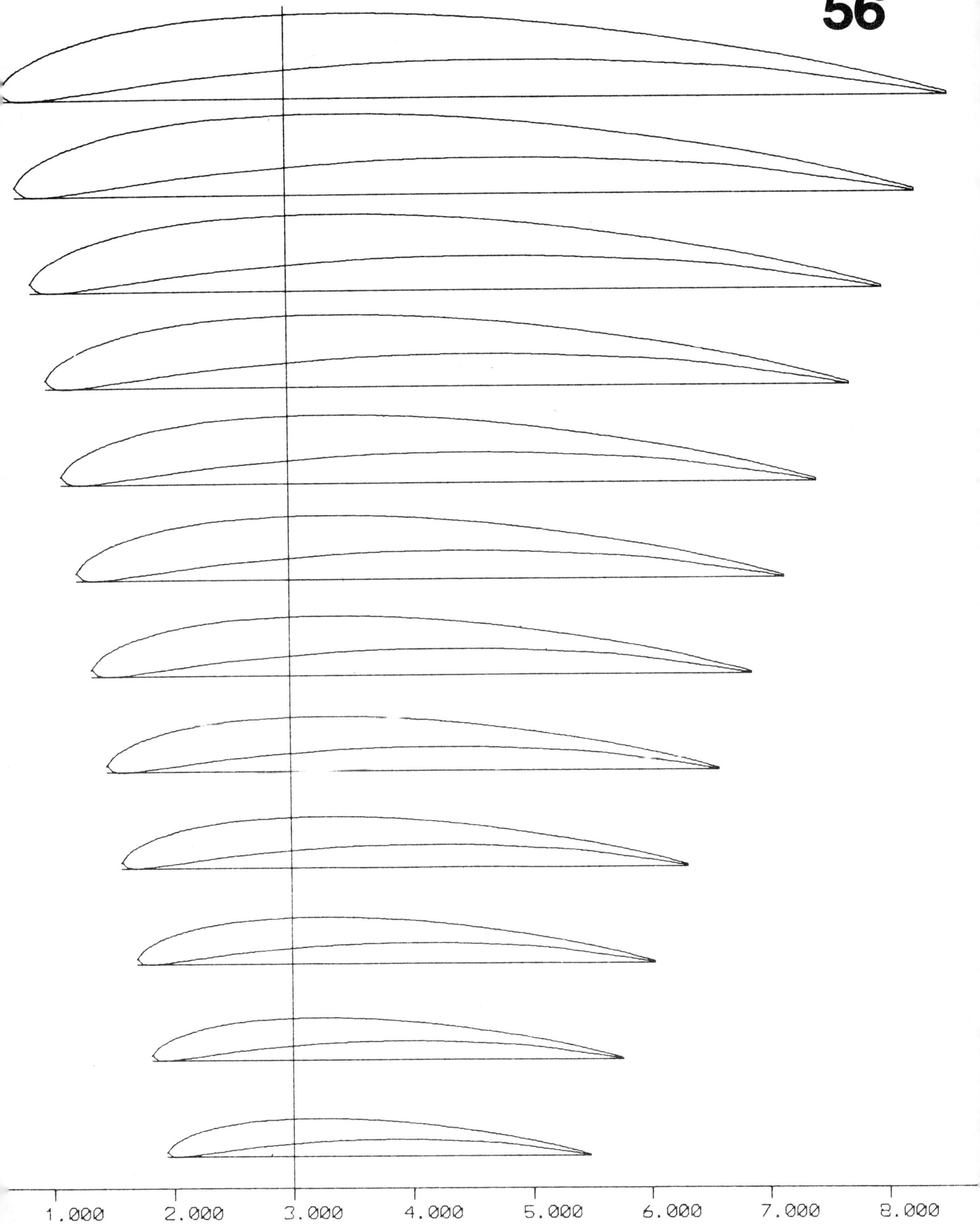

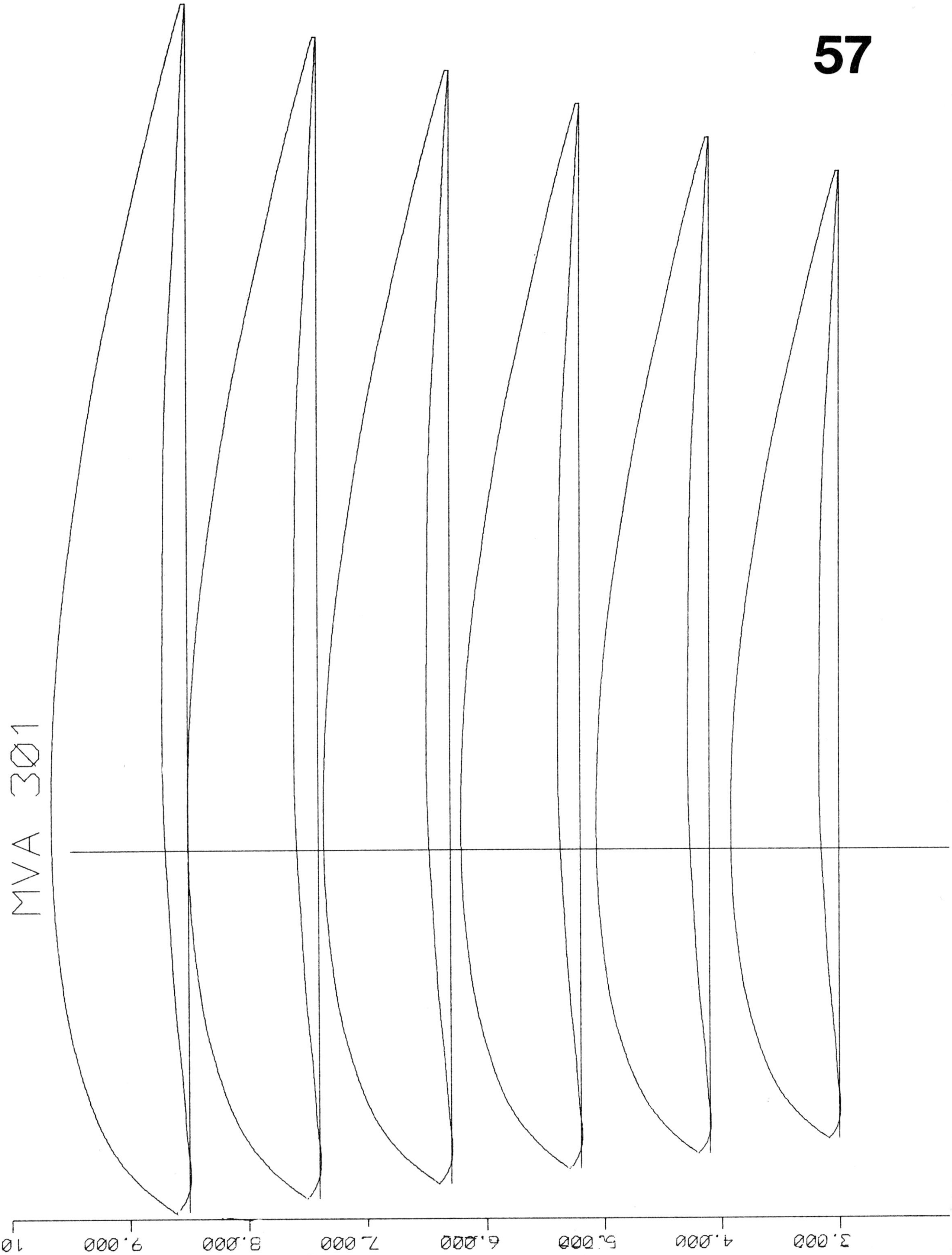
MVA 3Ø1
10
9.000
8.000
7.000
6.000
5.000
4.000
3.000

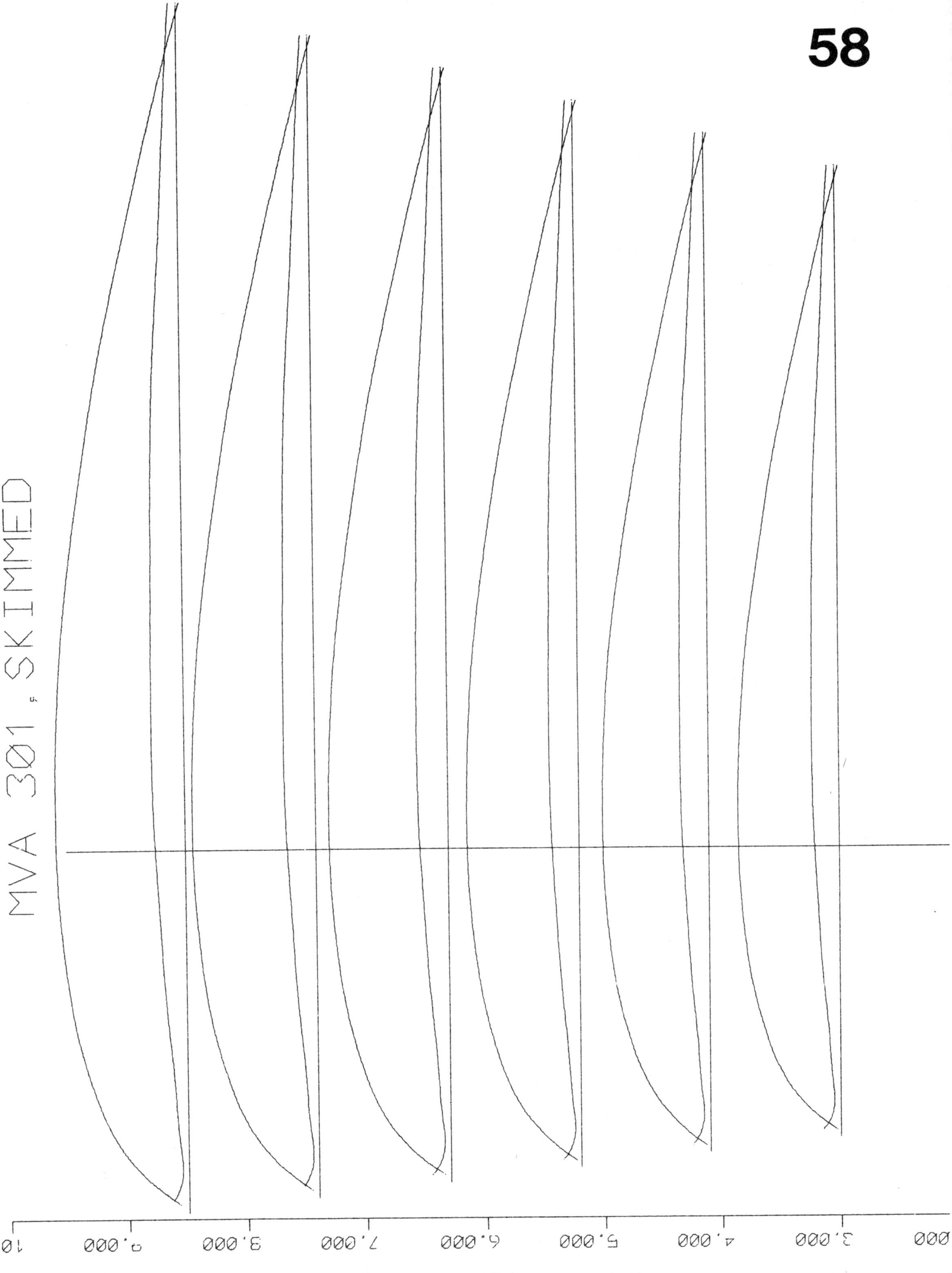
MVA 3Ø1, SKIMMED
10
9.000
8.000
7.000
6.000
5.000
4.000
3.000
000

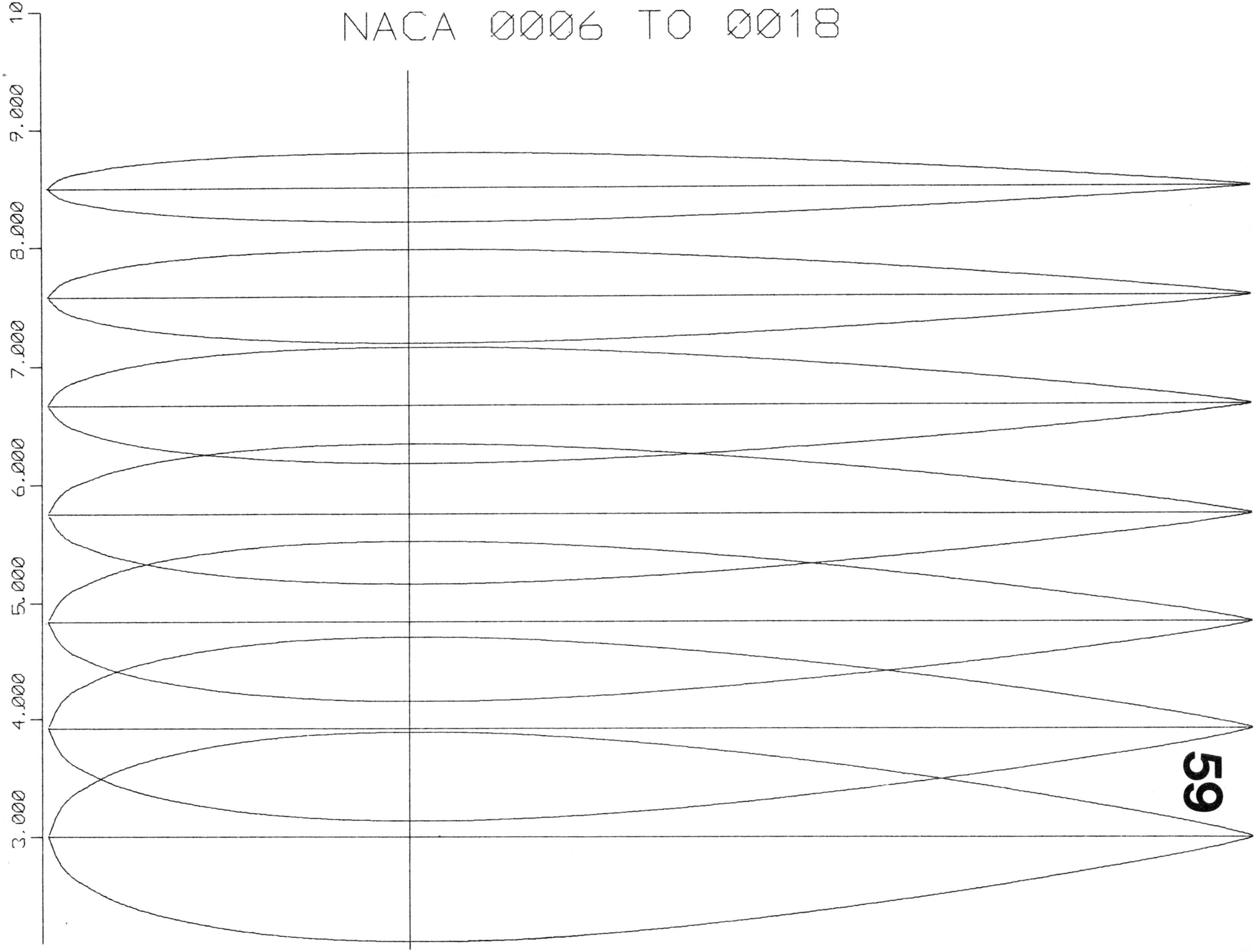
NACA 0006 TO 0018
10
9.000
8.000
7.000
6.000
5.000
4.000
3.000

NACA ØØØ6 TO ØØ18, SKIMMED

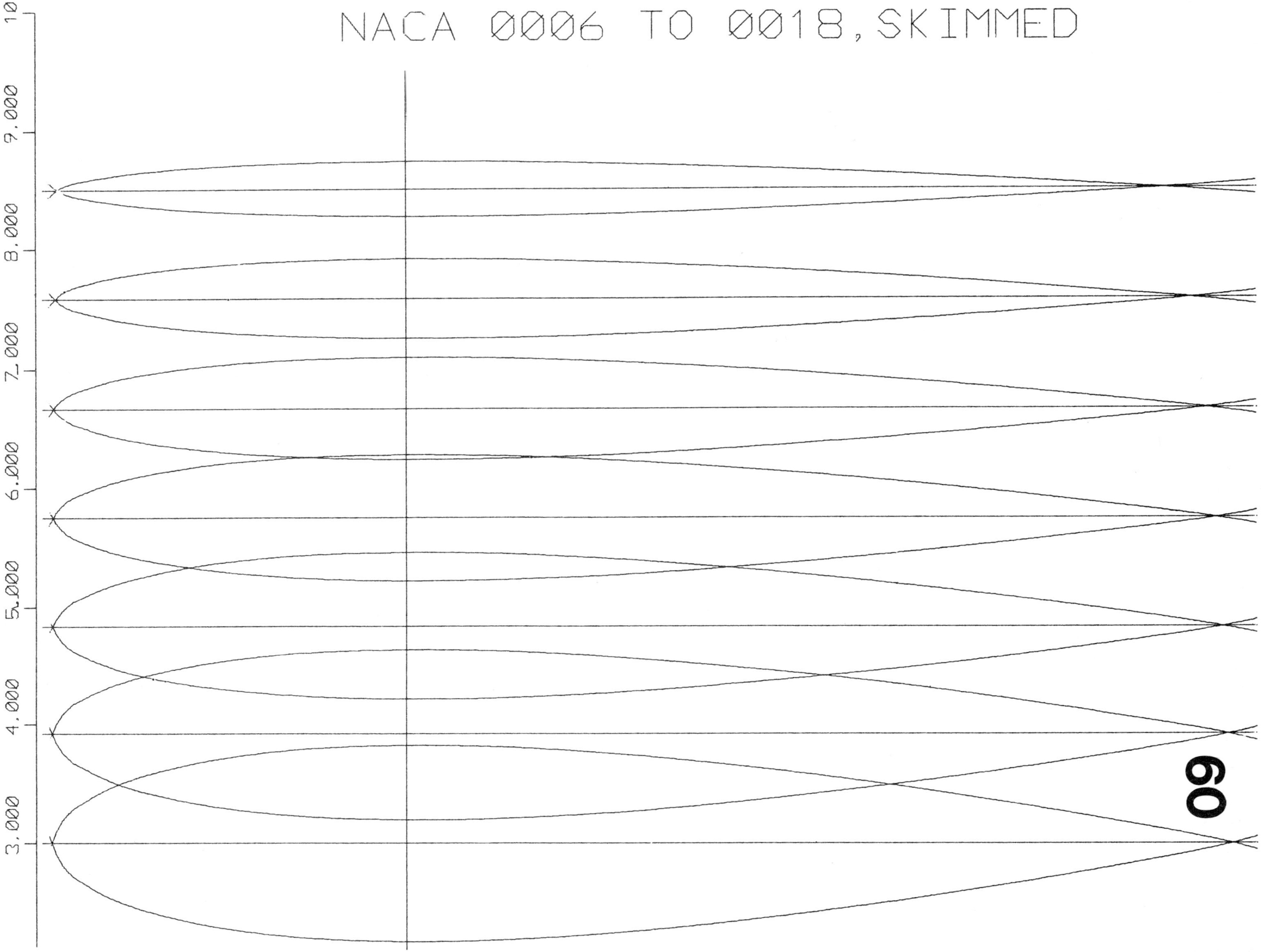

60

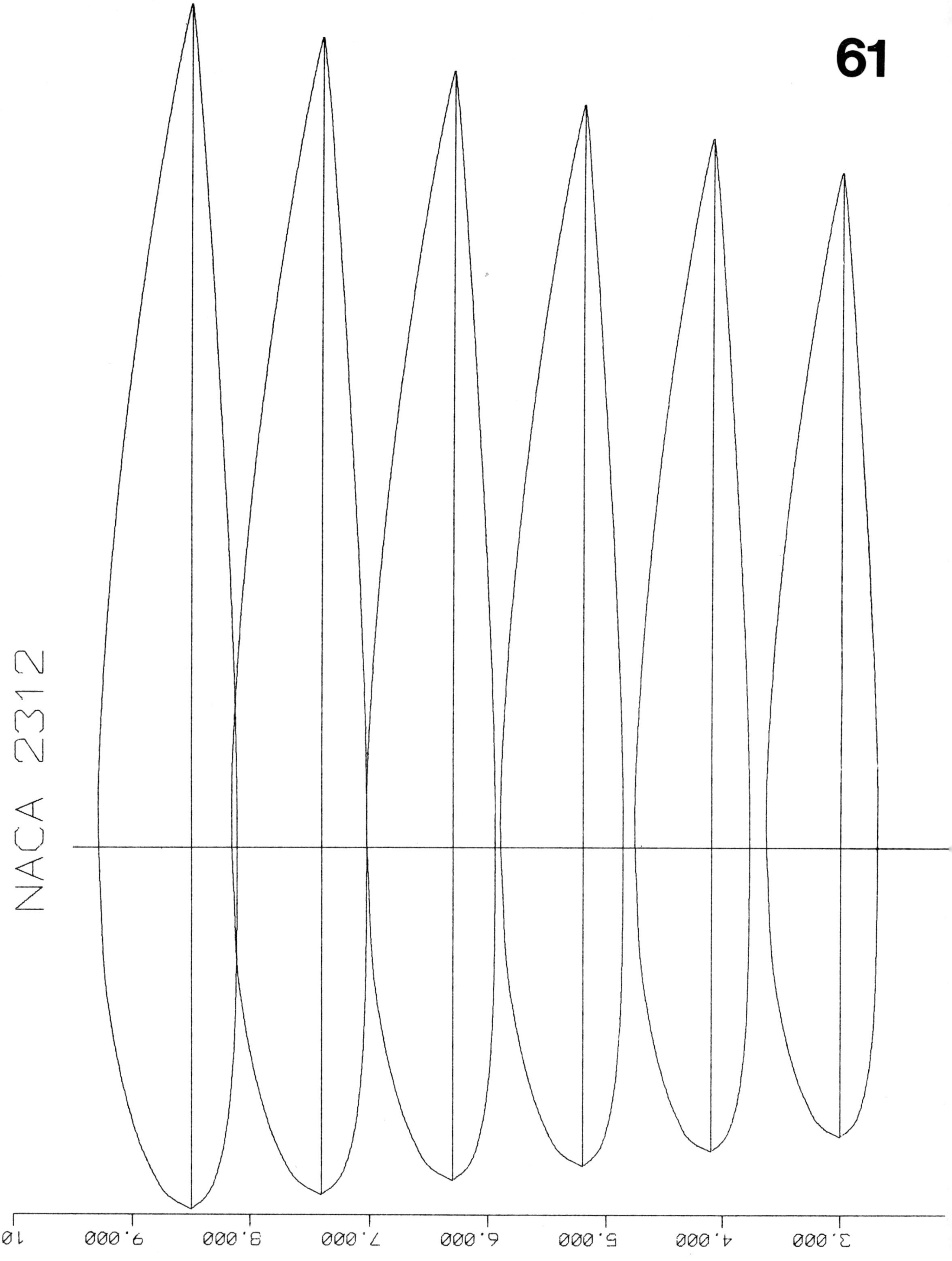
NACA 2312
10
9.000
8.000
7.000
6.000
5.000
4.000
3.000

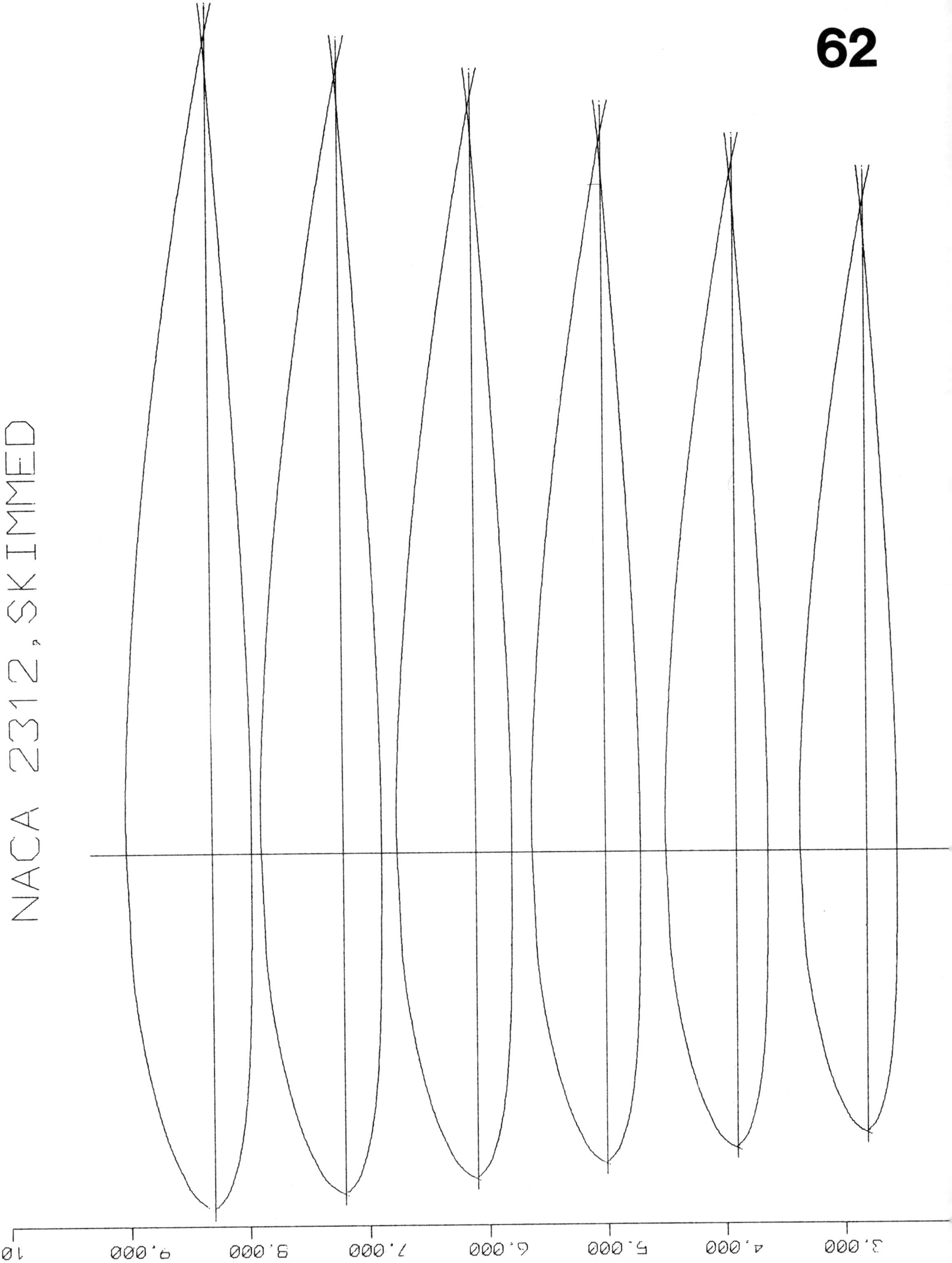
NACA 2312, SKIMMED
10
9.000
8.000
7.000
6.000
5.000
4.000
3.000

63

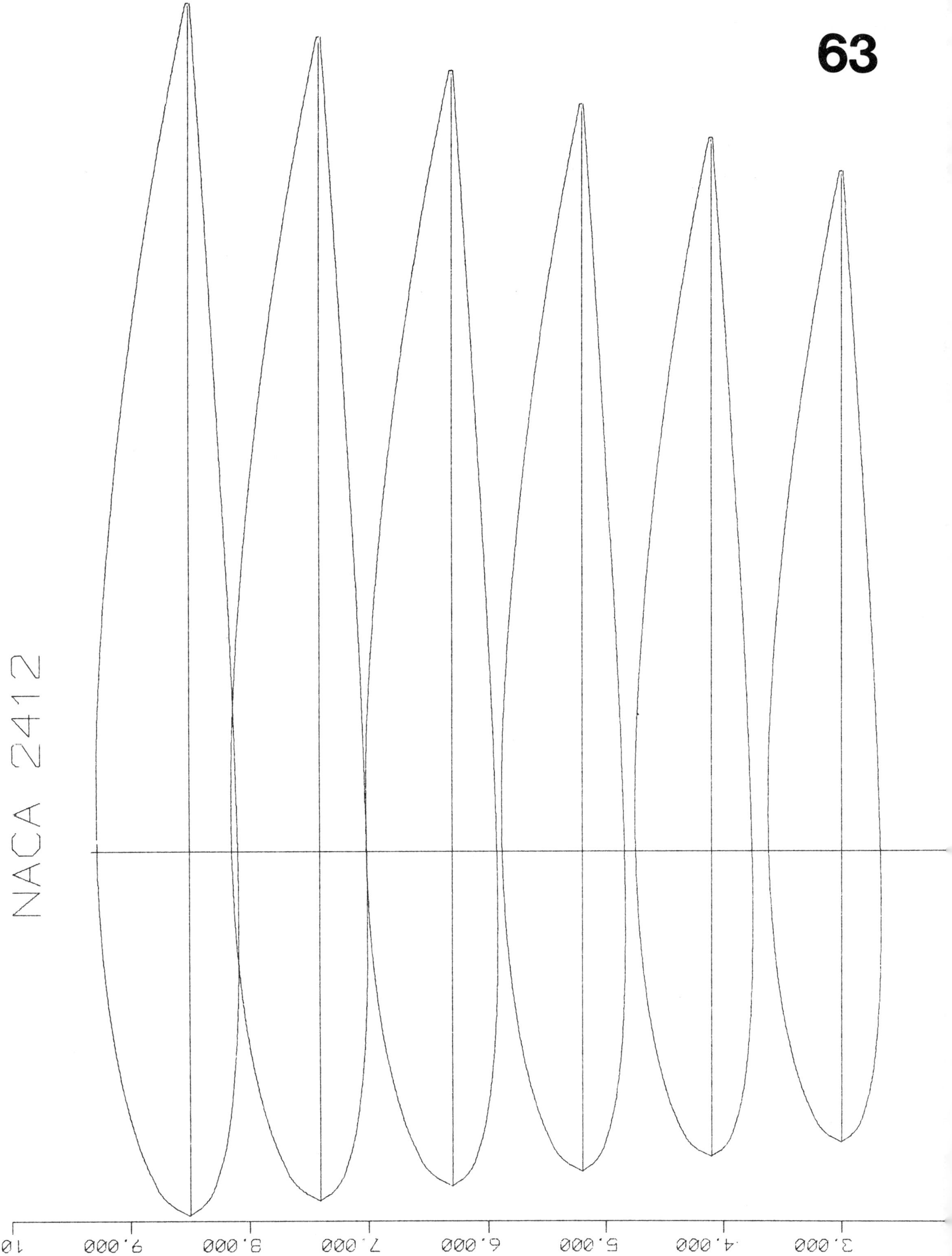

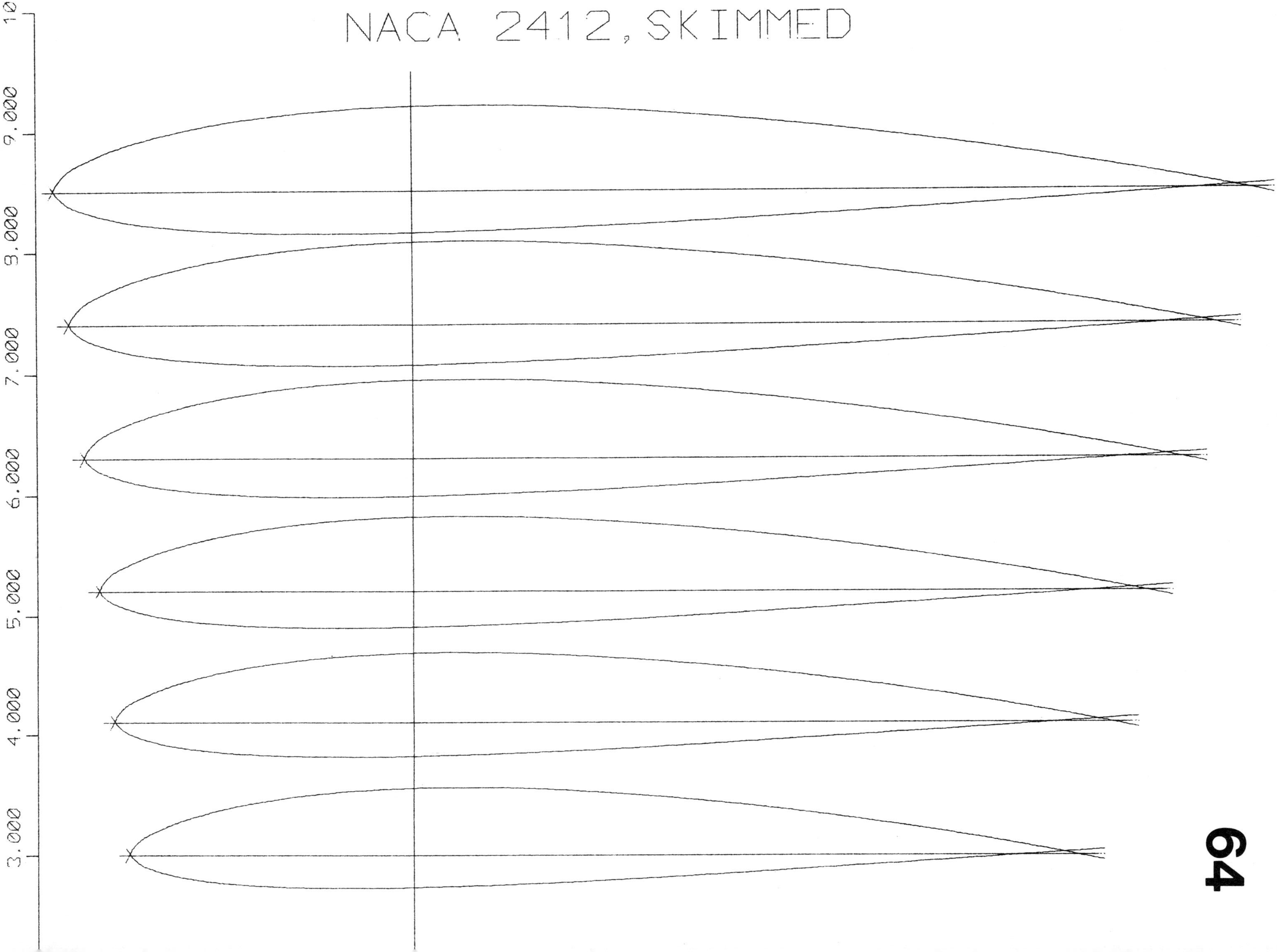

64

NACA 4409

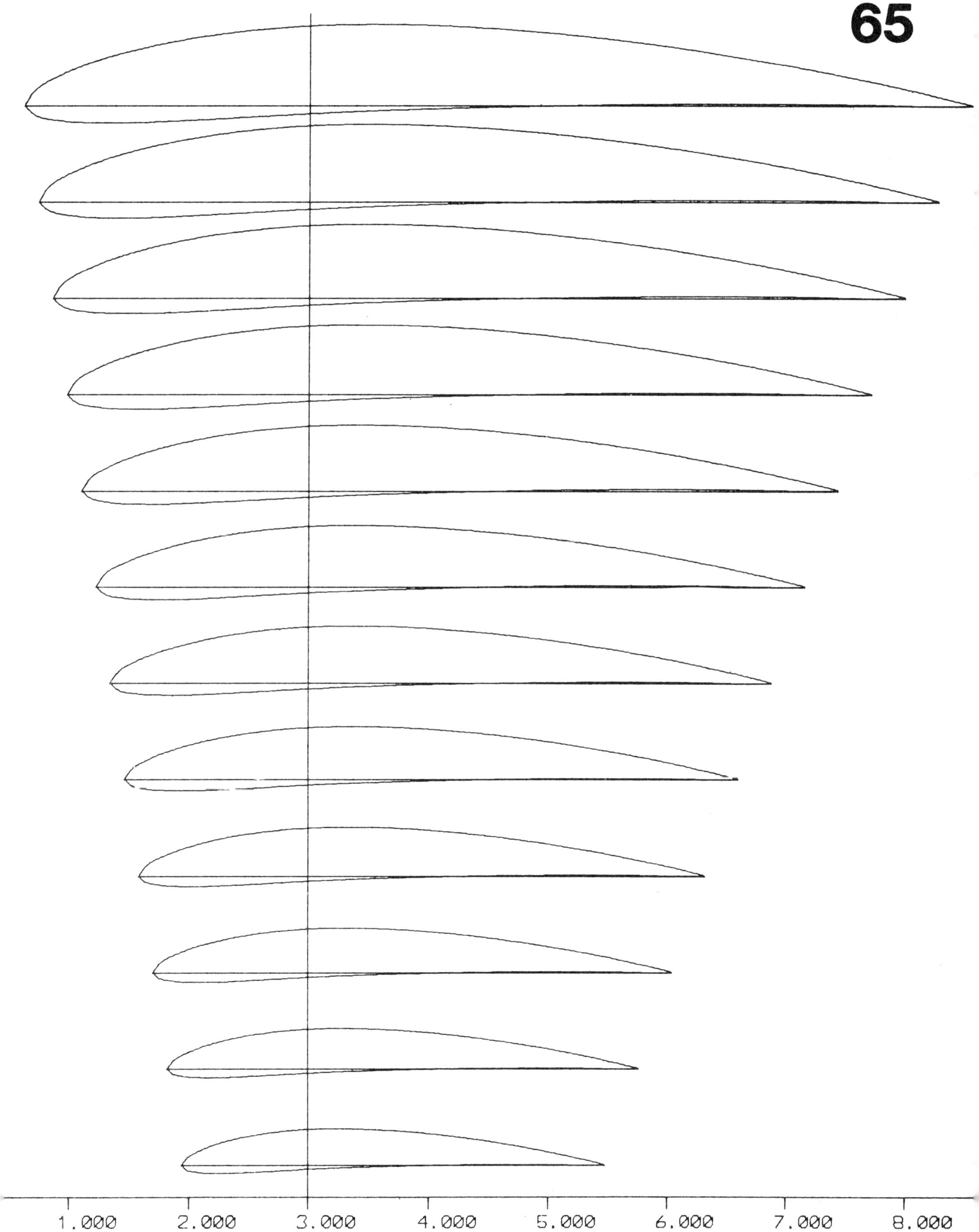

NACA 4409, SKIMMED

66

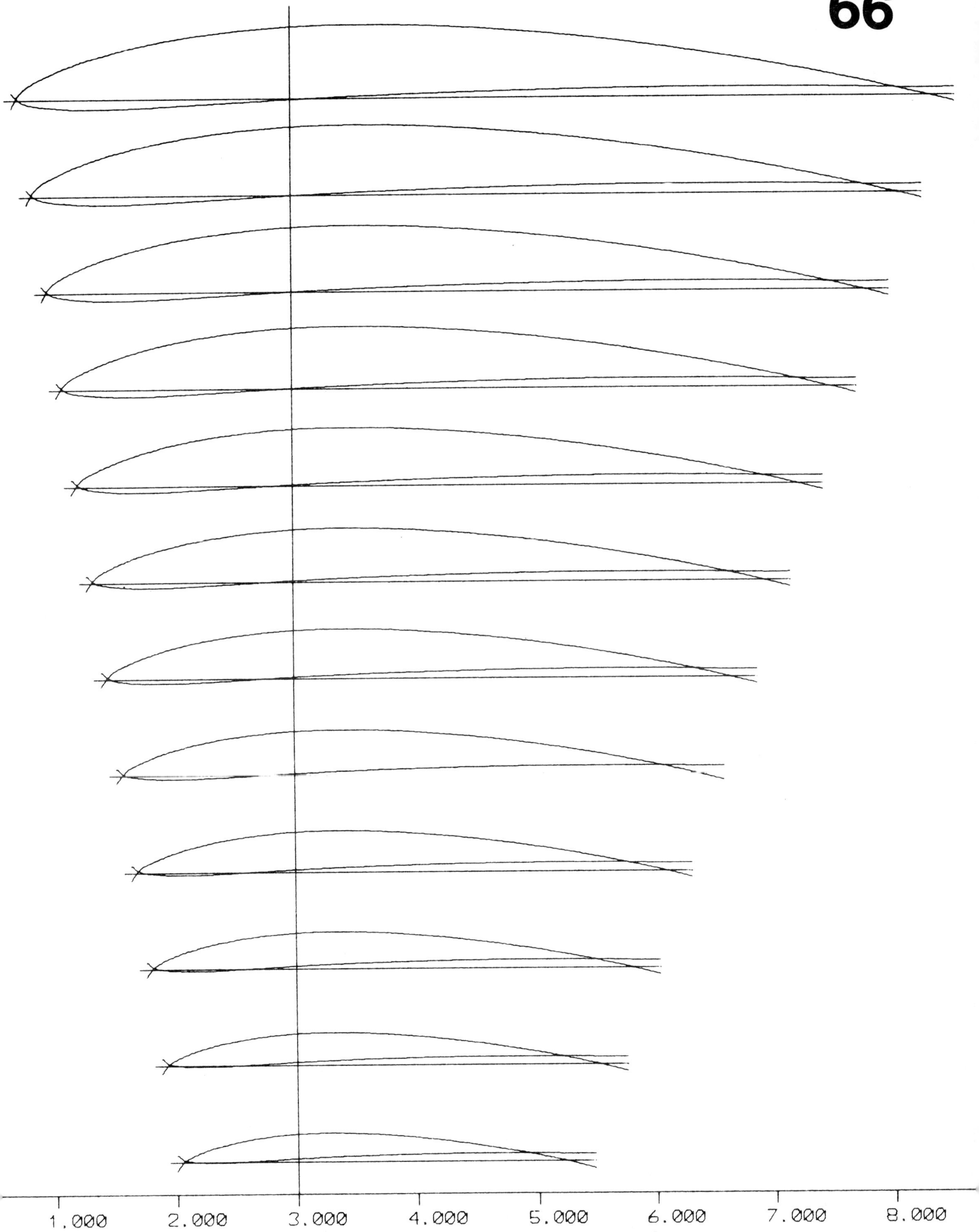

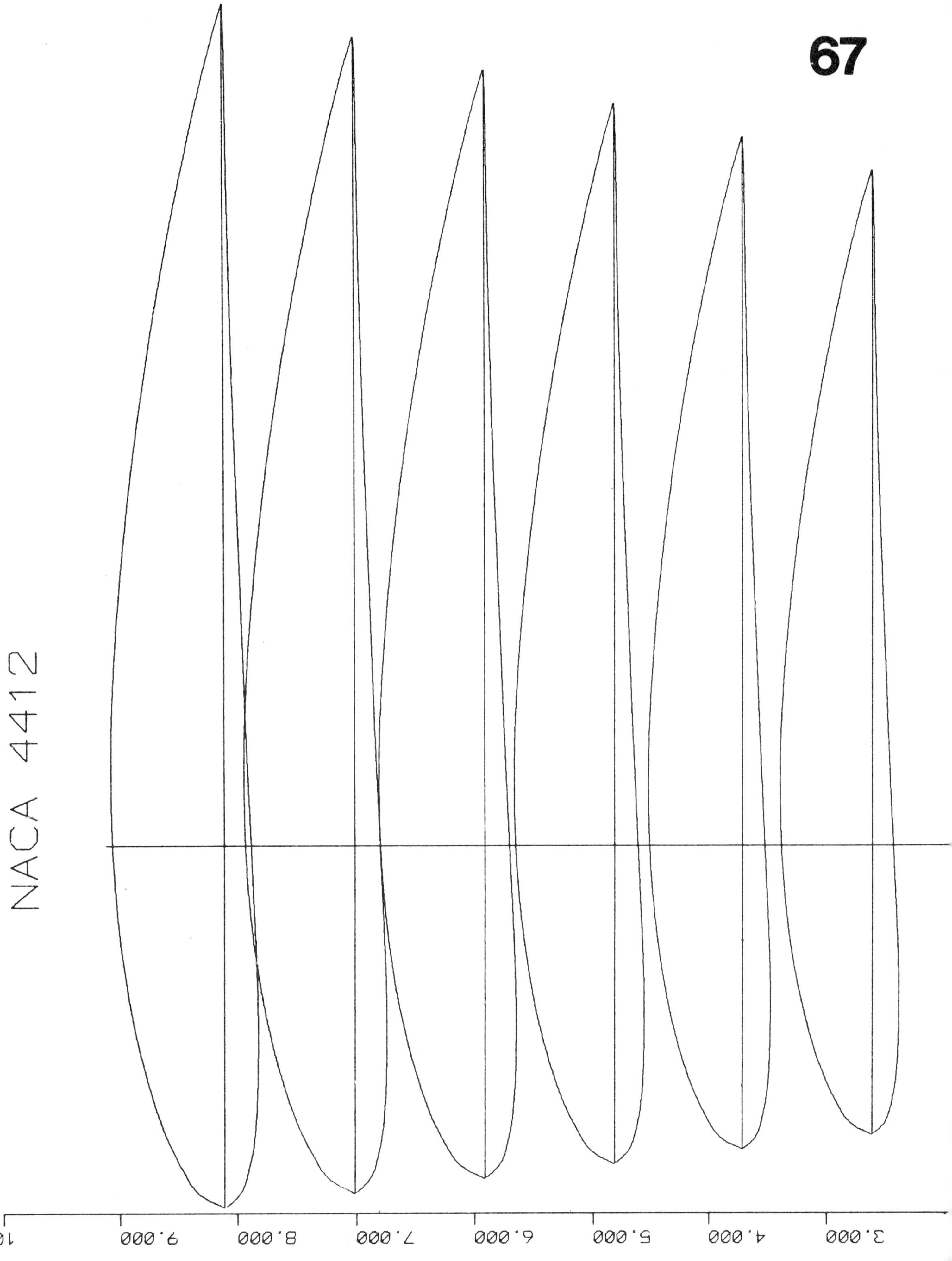
NACA 4412
10
9.000
8.000
7.000
6.000
5.000
4.000
3.000

NACA 4412, SKIMMED

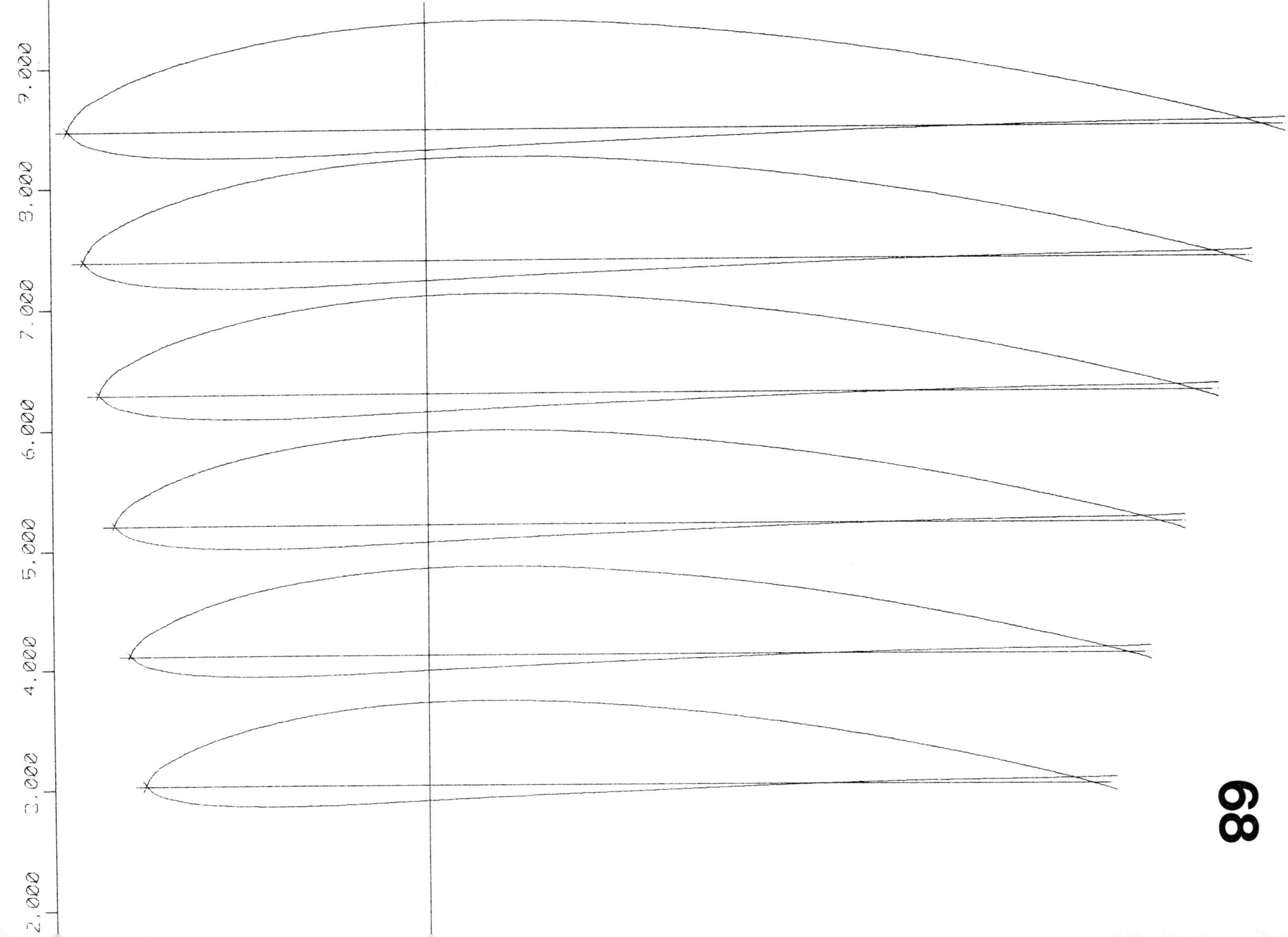

NACA 6409

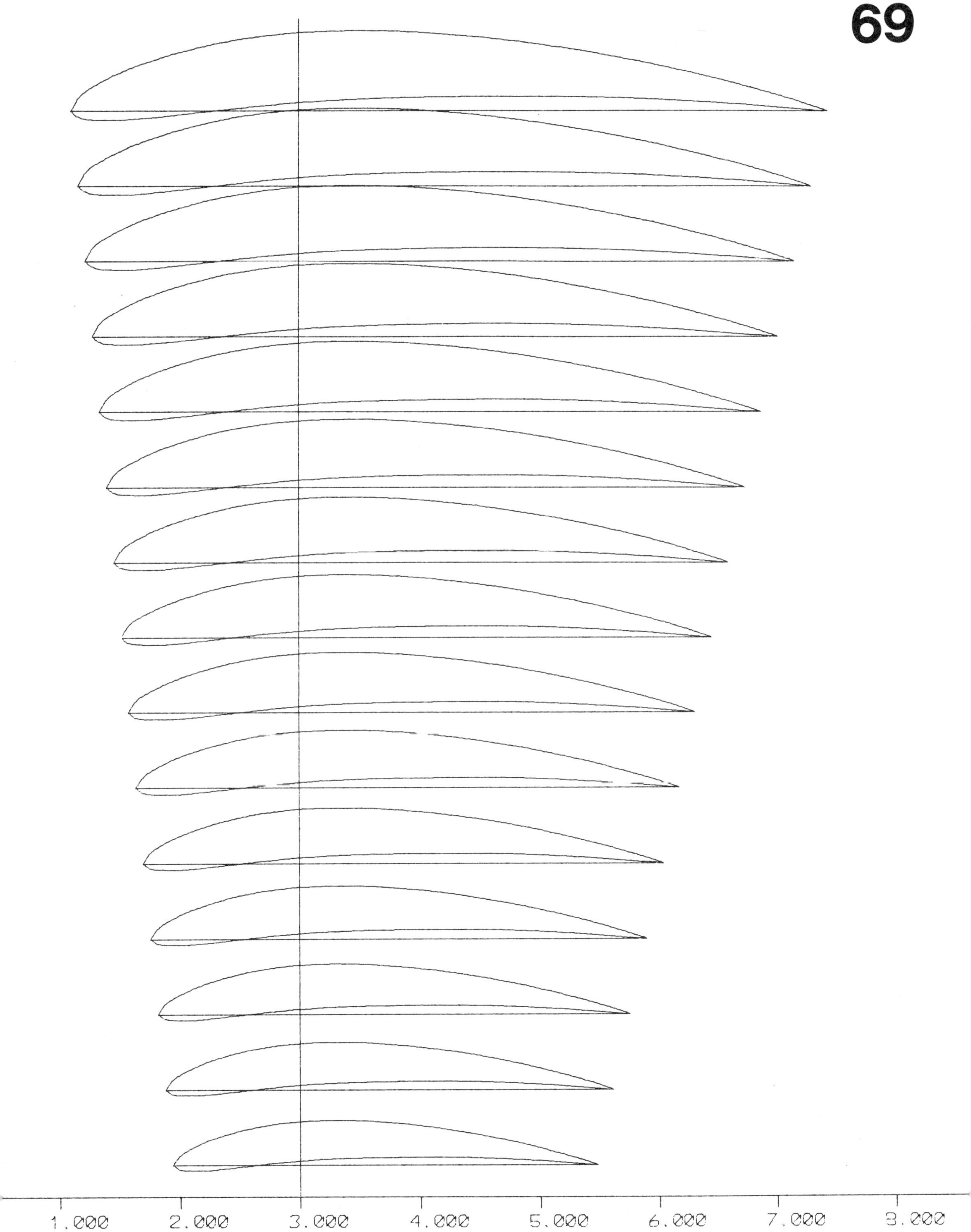

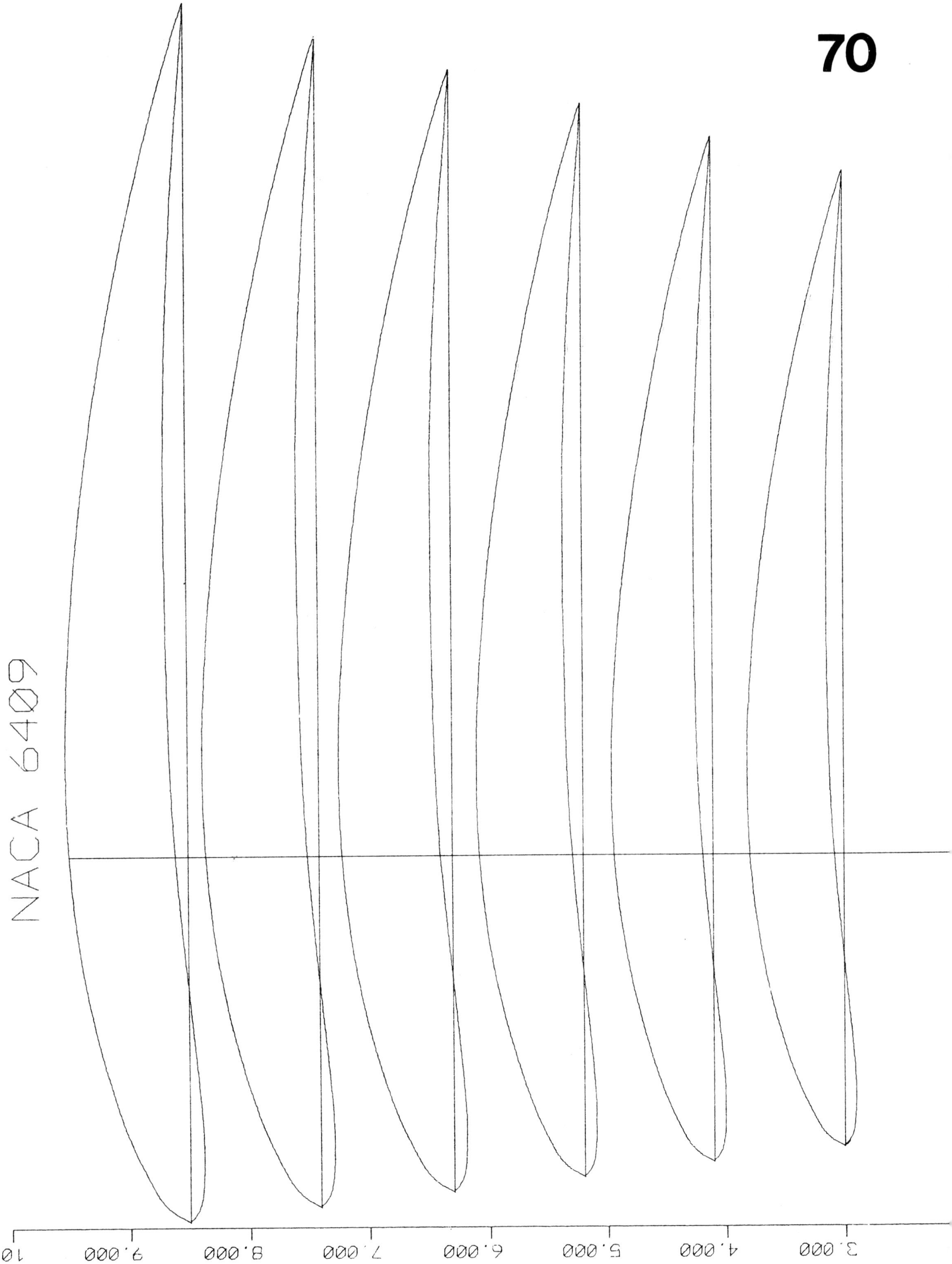
NACA 6409
10
9.000
8.000
7.000
6.000
5.000
4.000
3.000

65

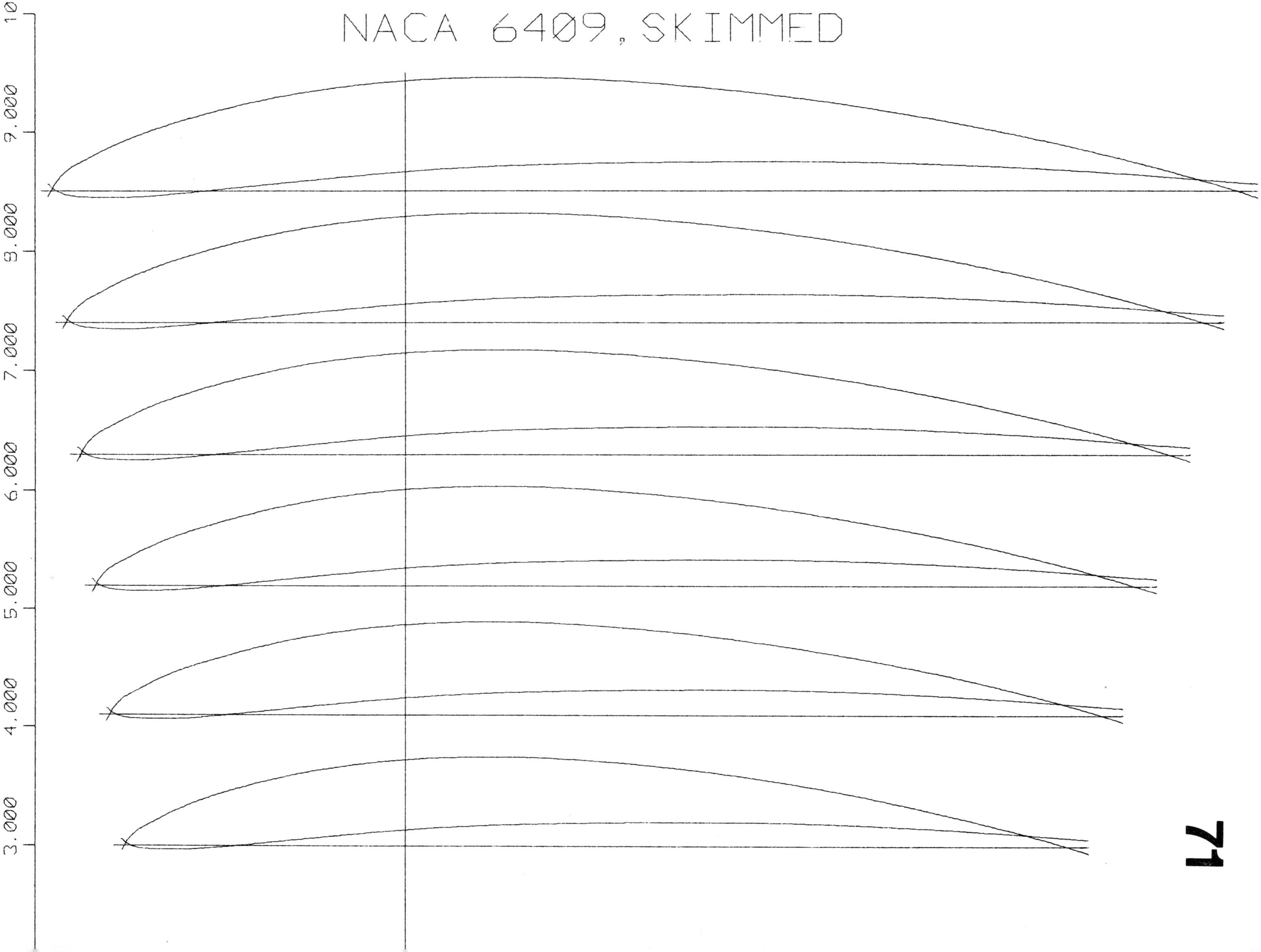

NACA 6409, SKIMMED
10
9.000
8.000
7.000
6.000
5.000
4.000
3.000

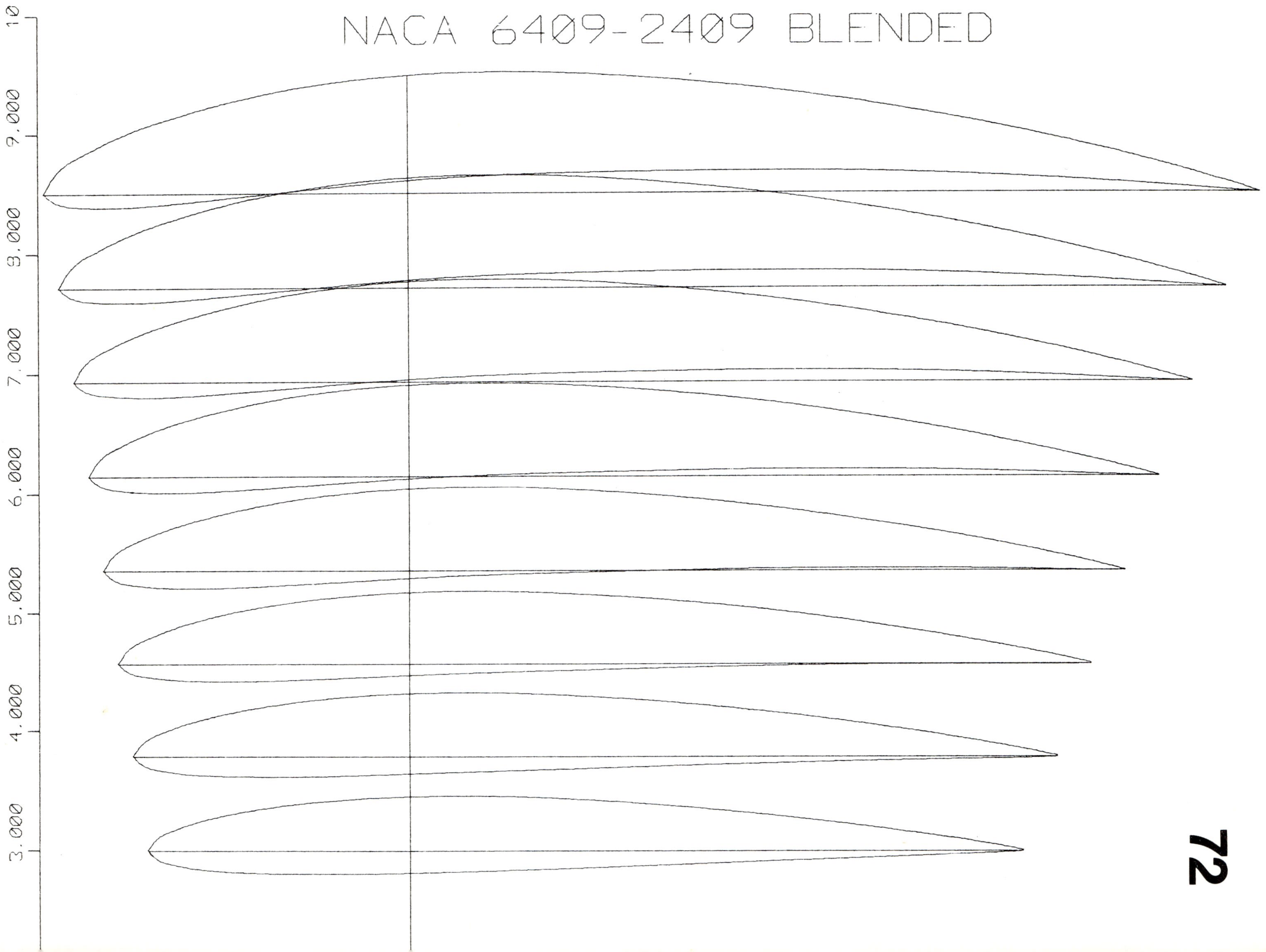
NACA 6409-2409 BLENDED
10
9.000
8.000
7.000
6.000
5.000
4.000
3.000

NACA 6409-2409 BLENDED, SKIMMED

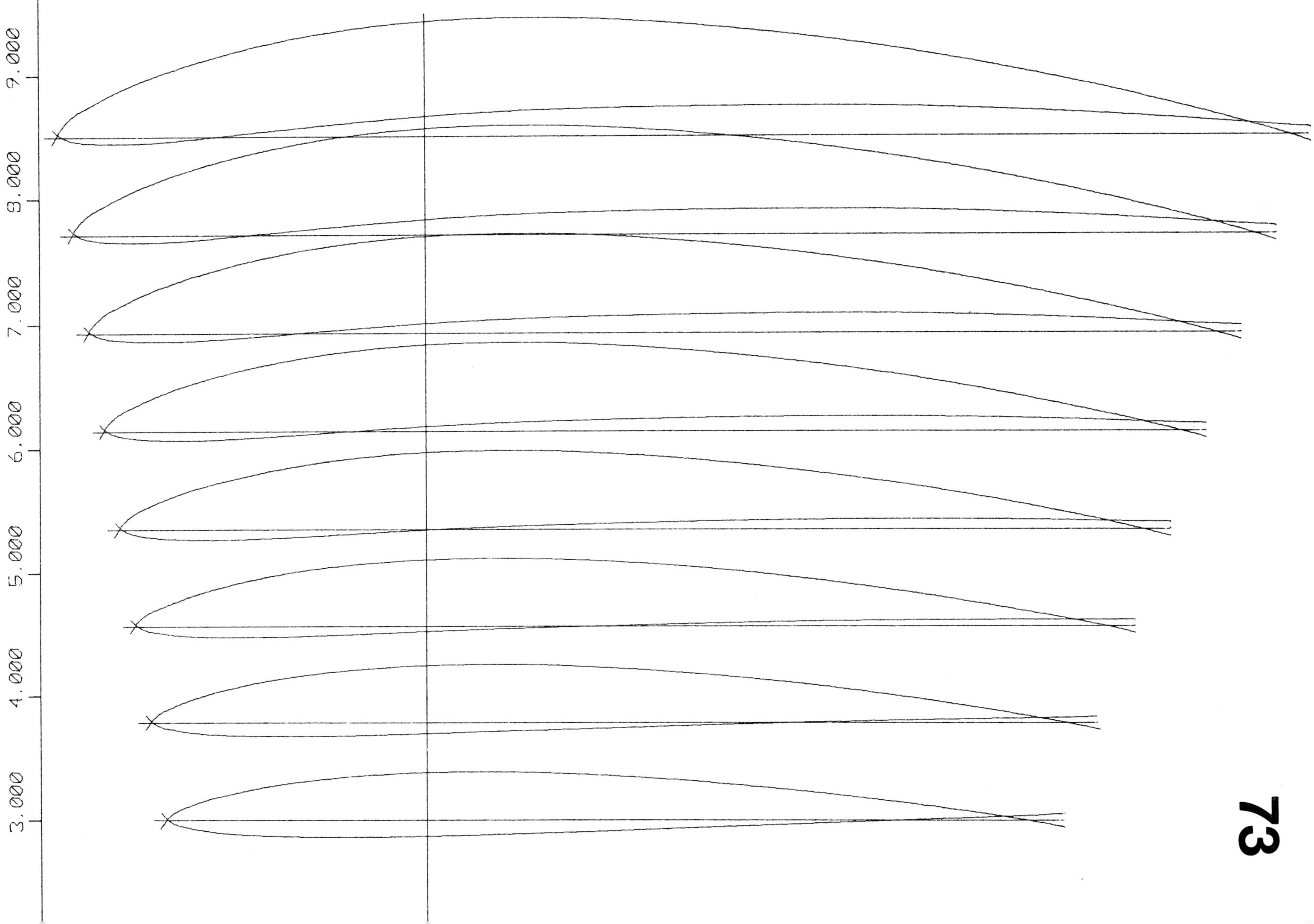

NEELMEYER

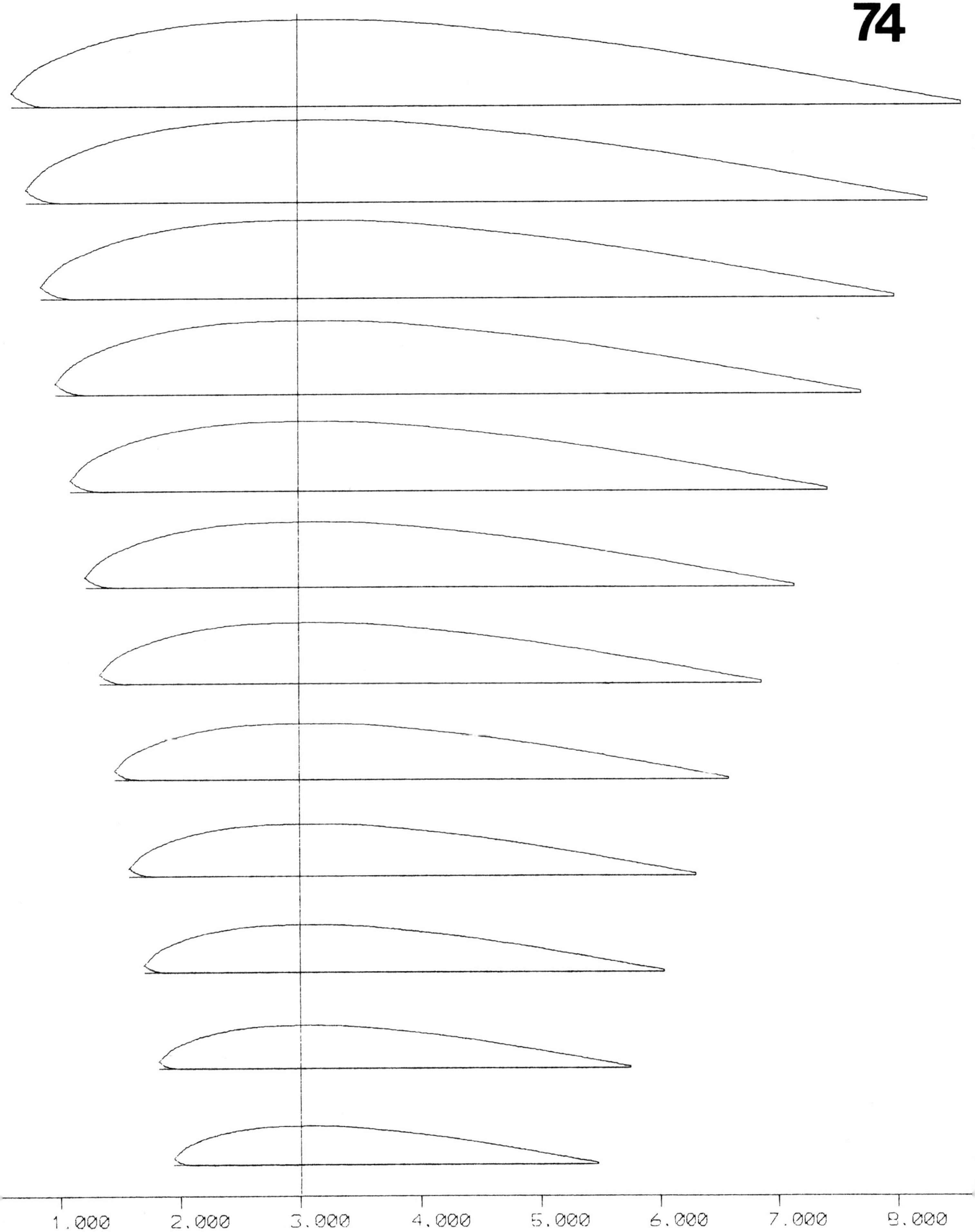

NEELMEYER, SKIMMED

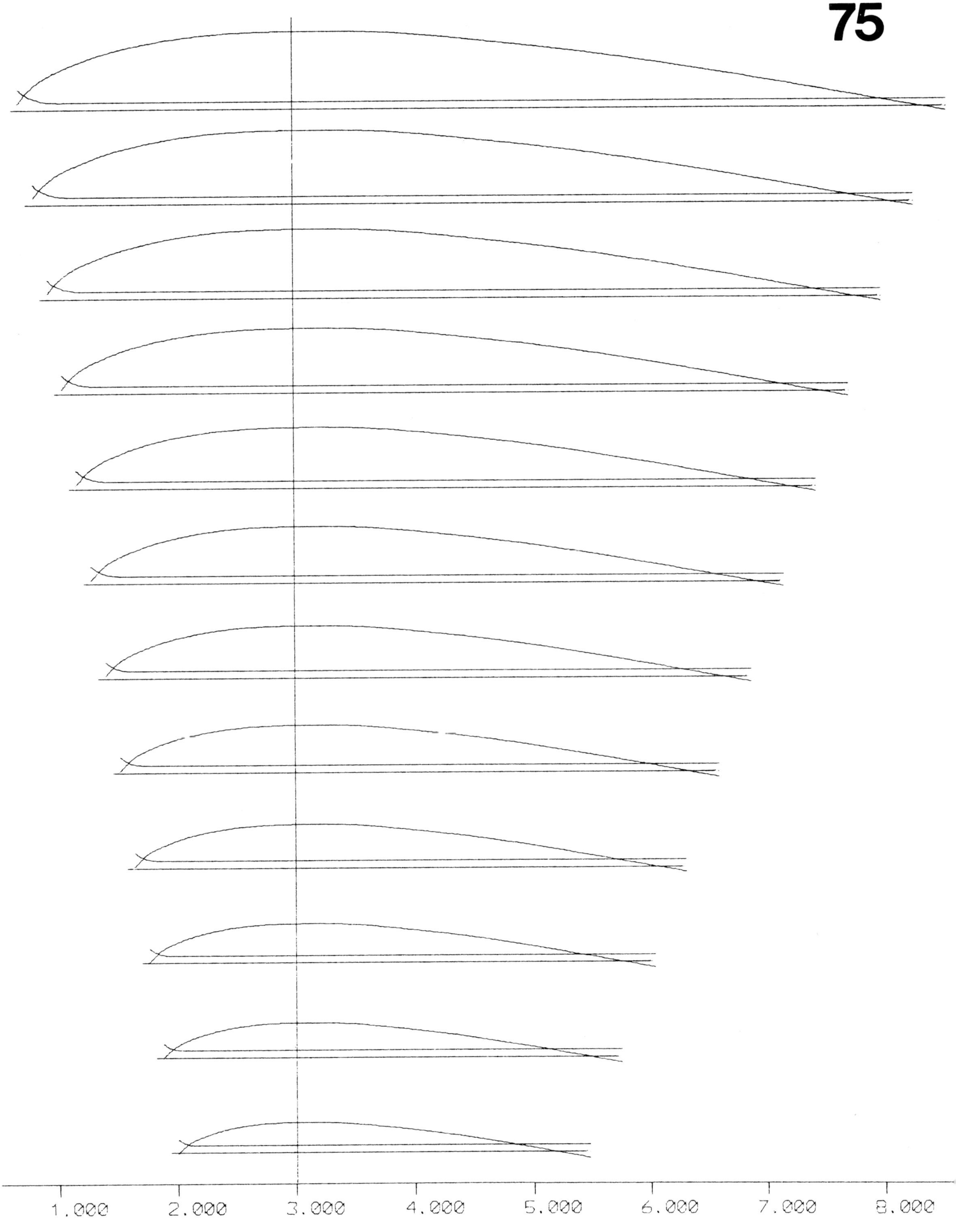

RITZ 7.45-5.5G

76

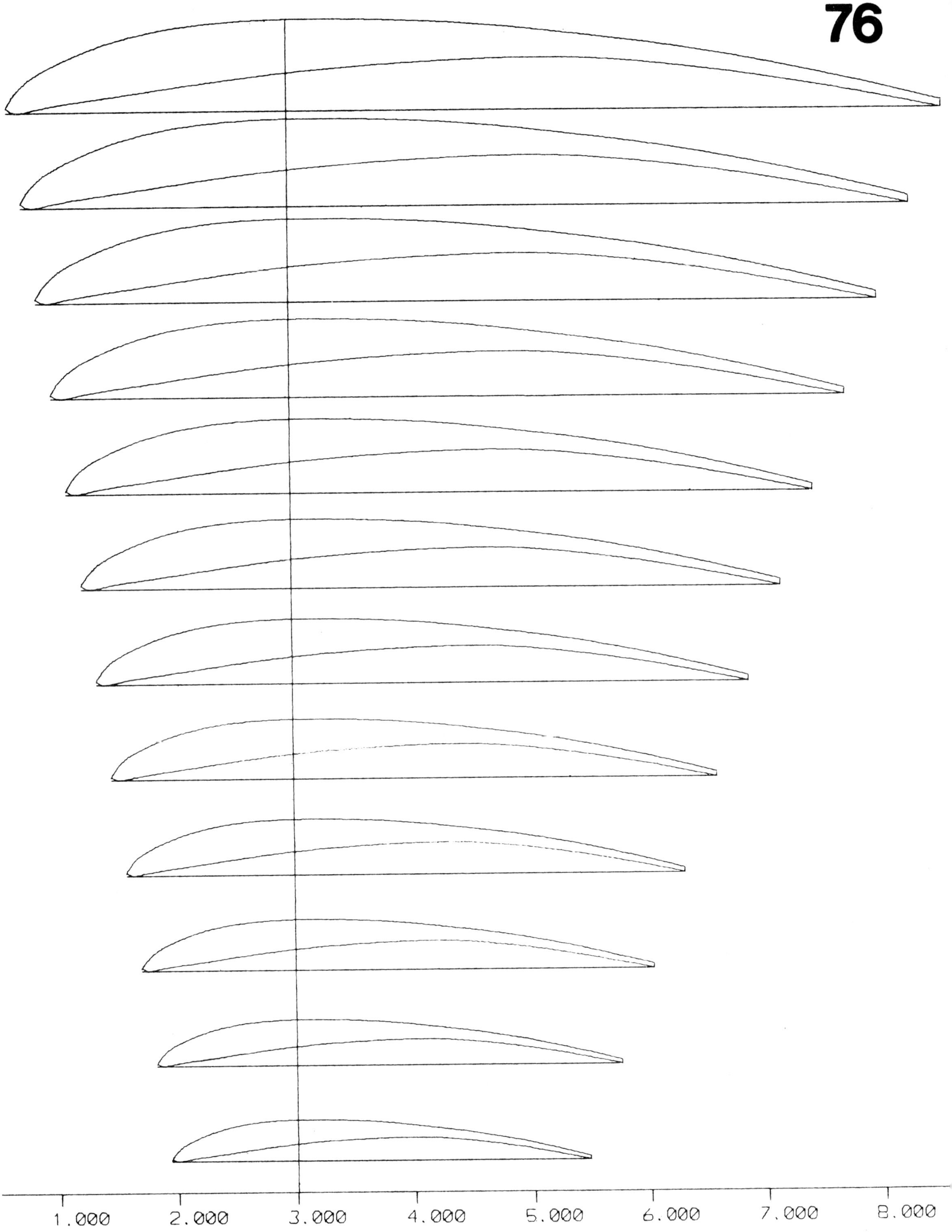

SCHWARTZBACH 68

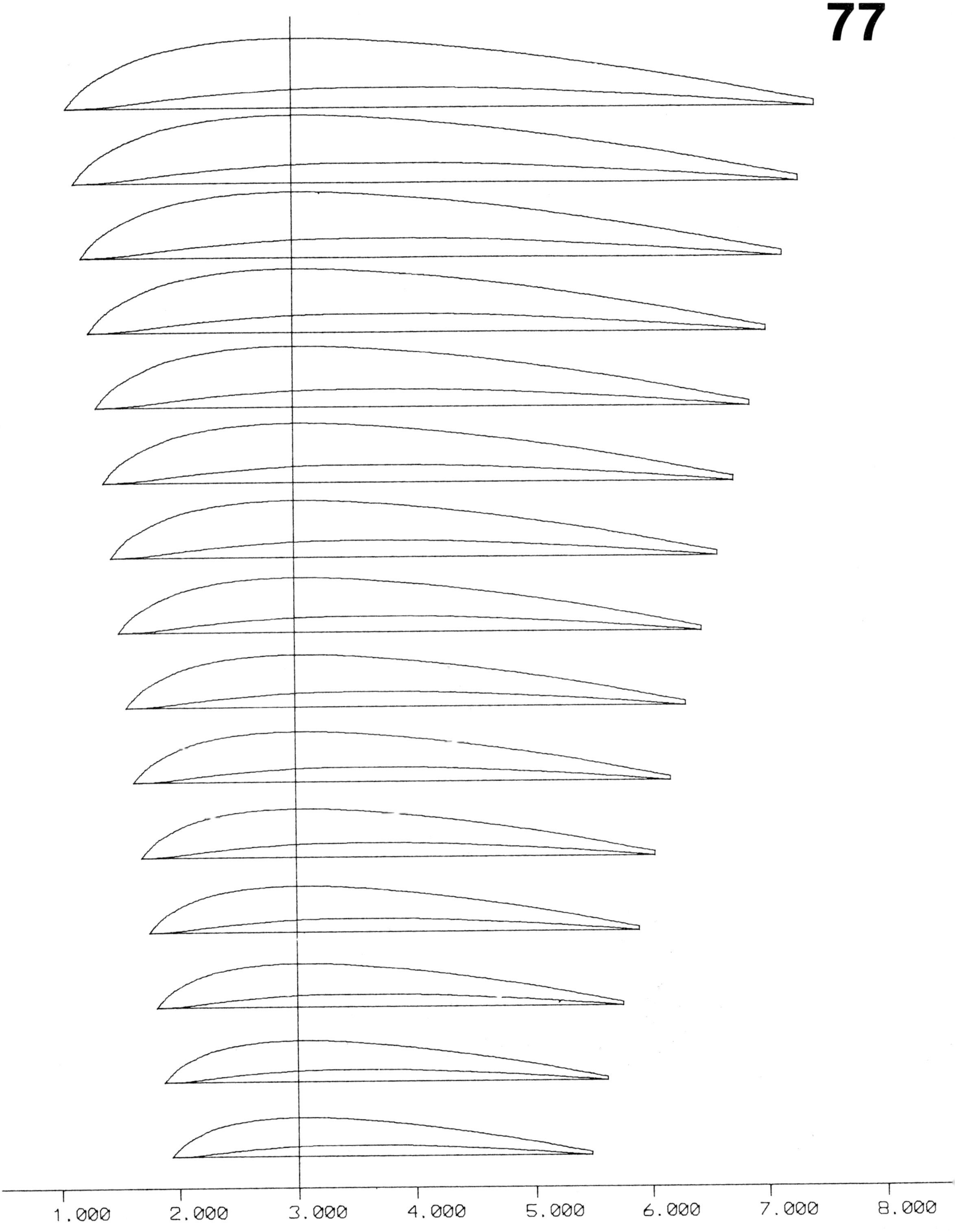

SCHWARTZBACH 68, SKIMMED

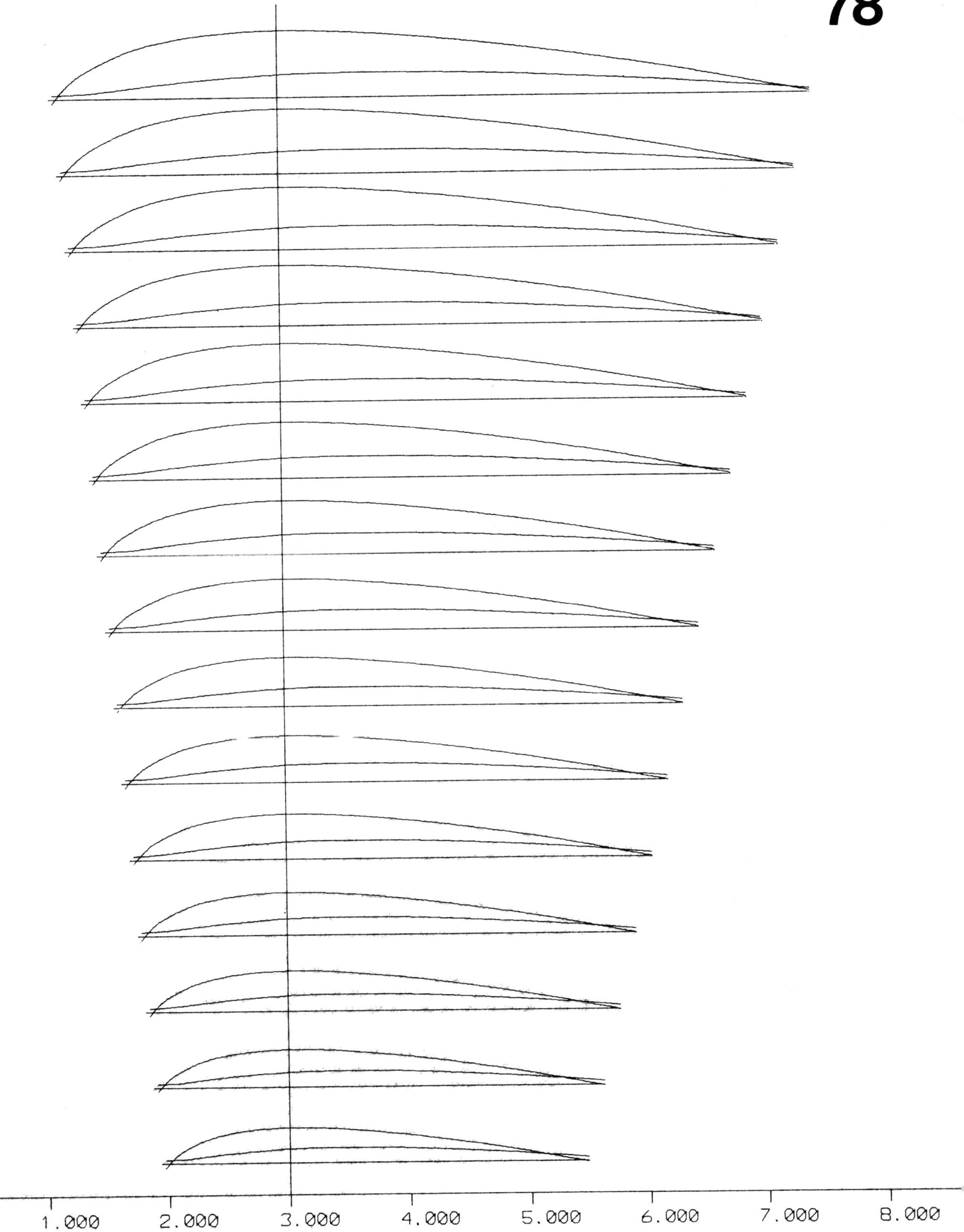

SCHWEINSBERG

79

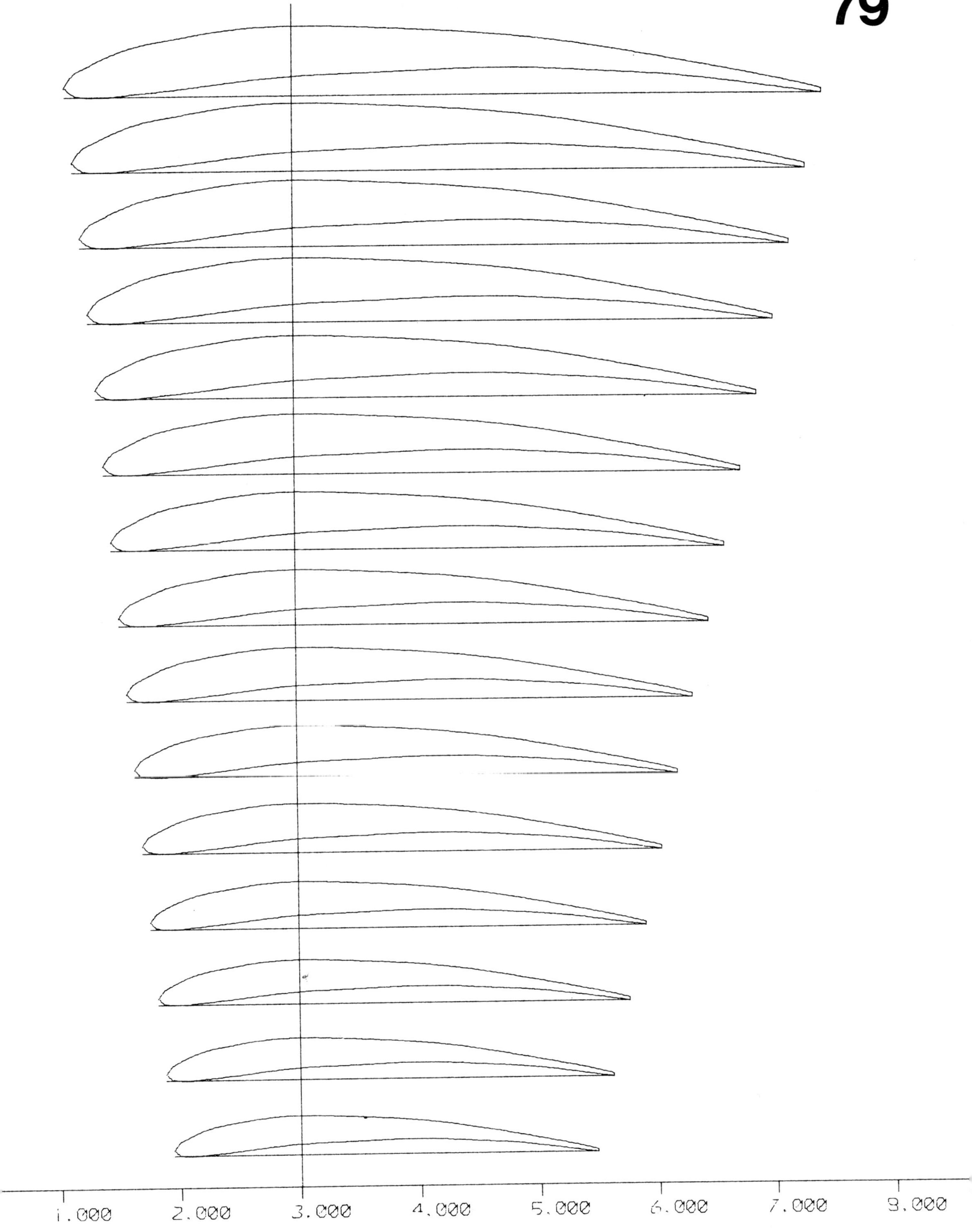

SCHWEINSBERG, SKIMMED

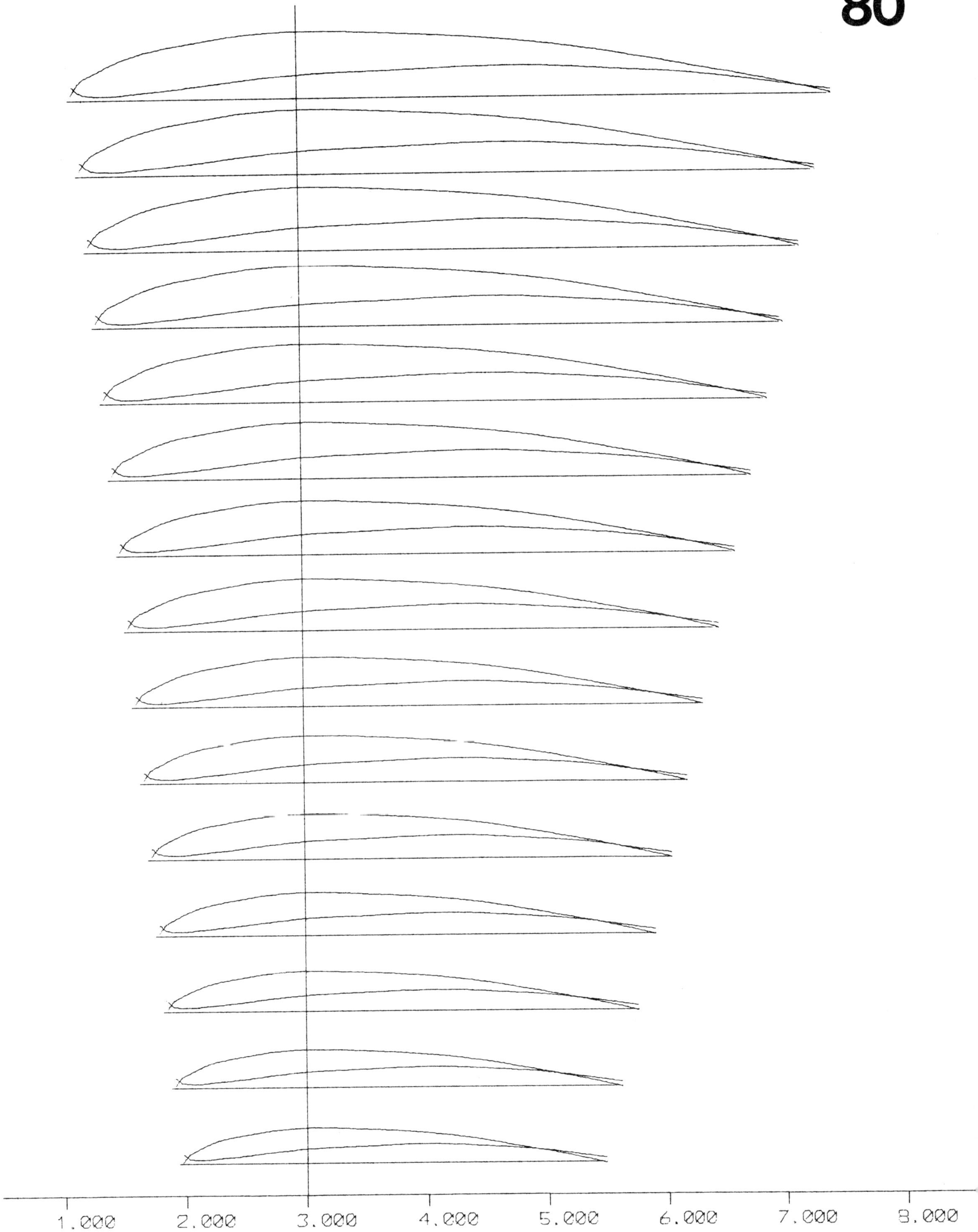

THOMANN F4

81

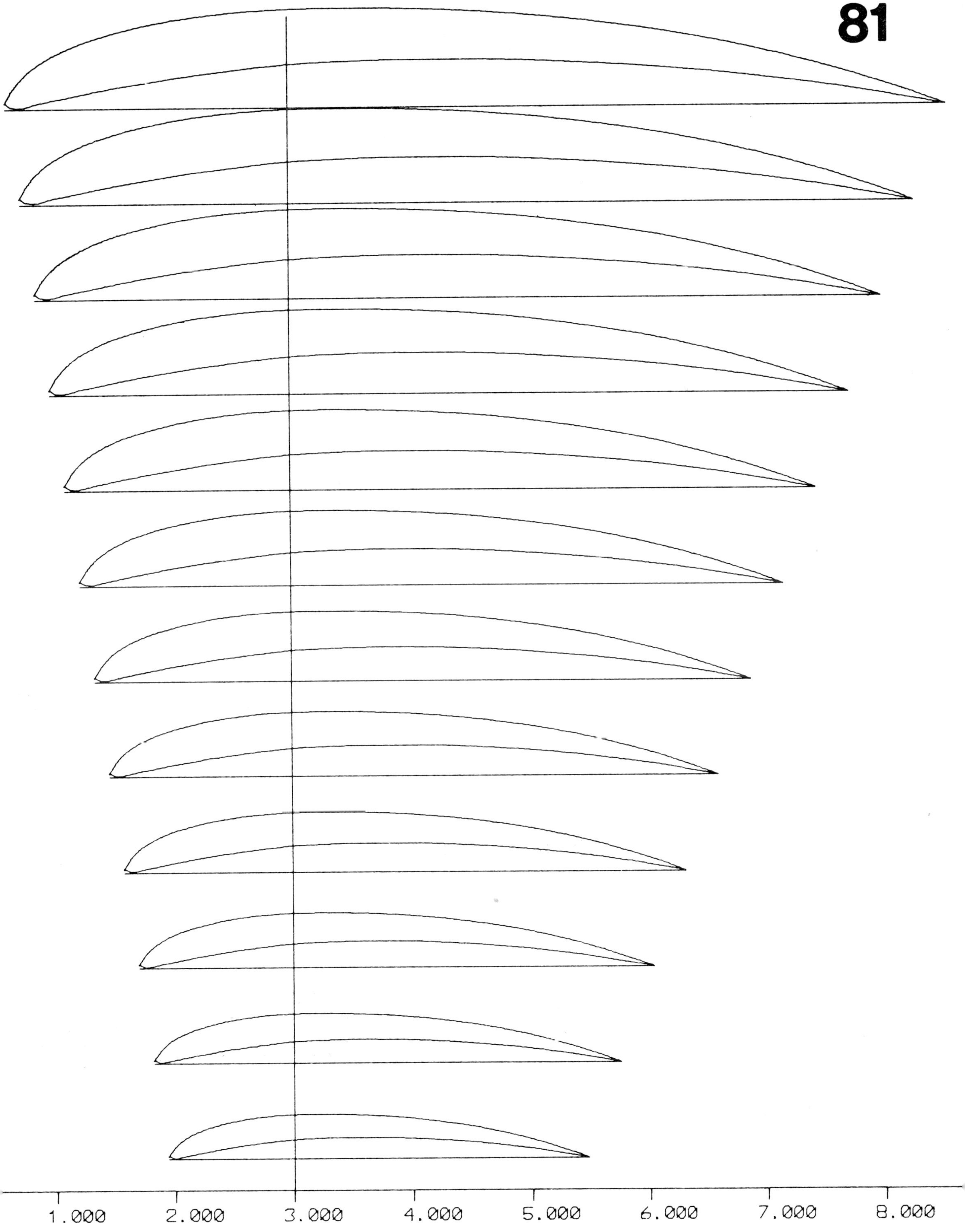

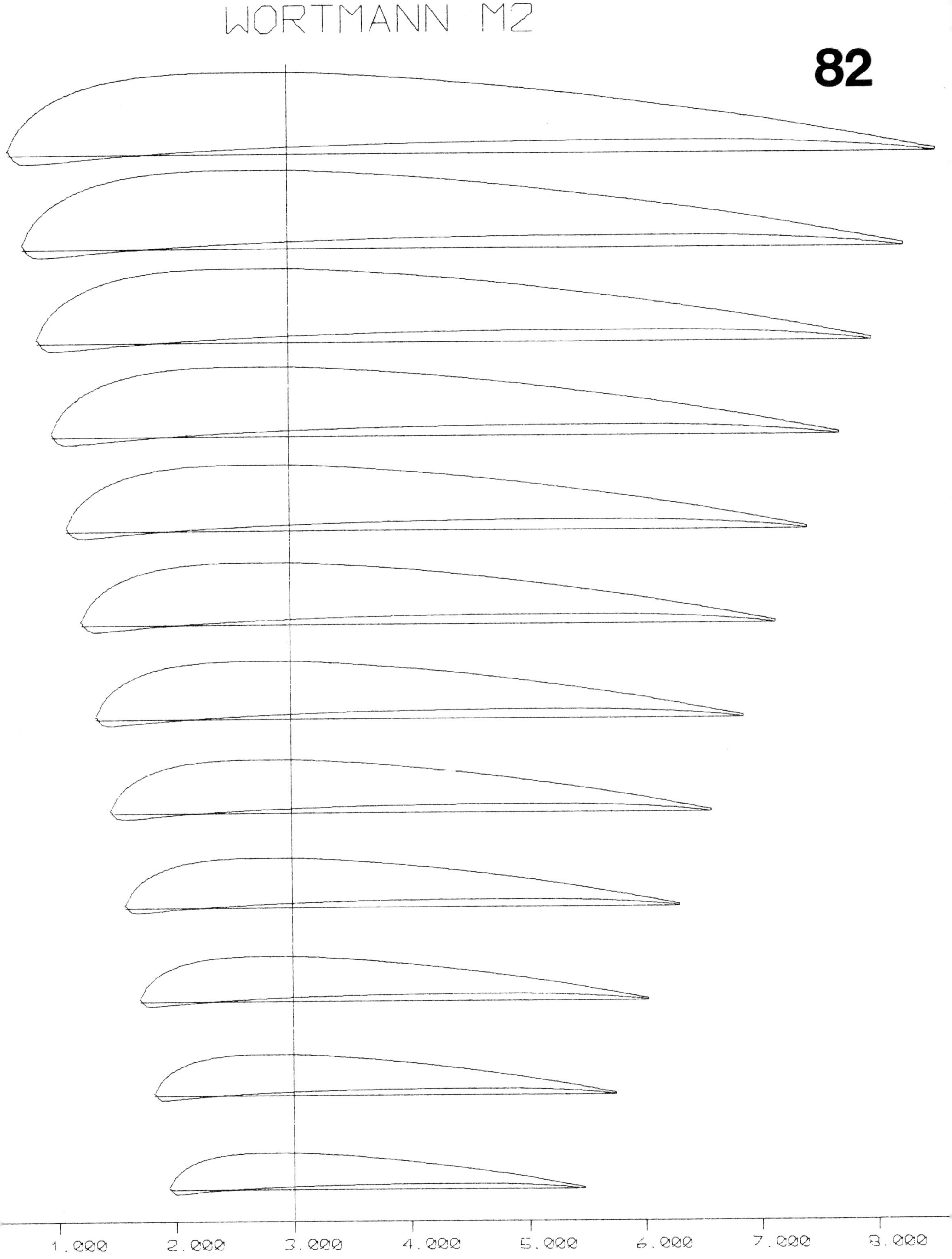
WORTMANN M2
1.000
2.000
3.000
4.000
5.000
6.000
7.000
8.000

WORTMANN M2, SKIMMED

83

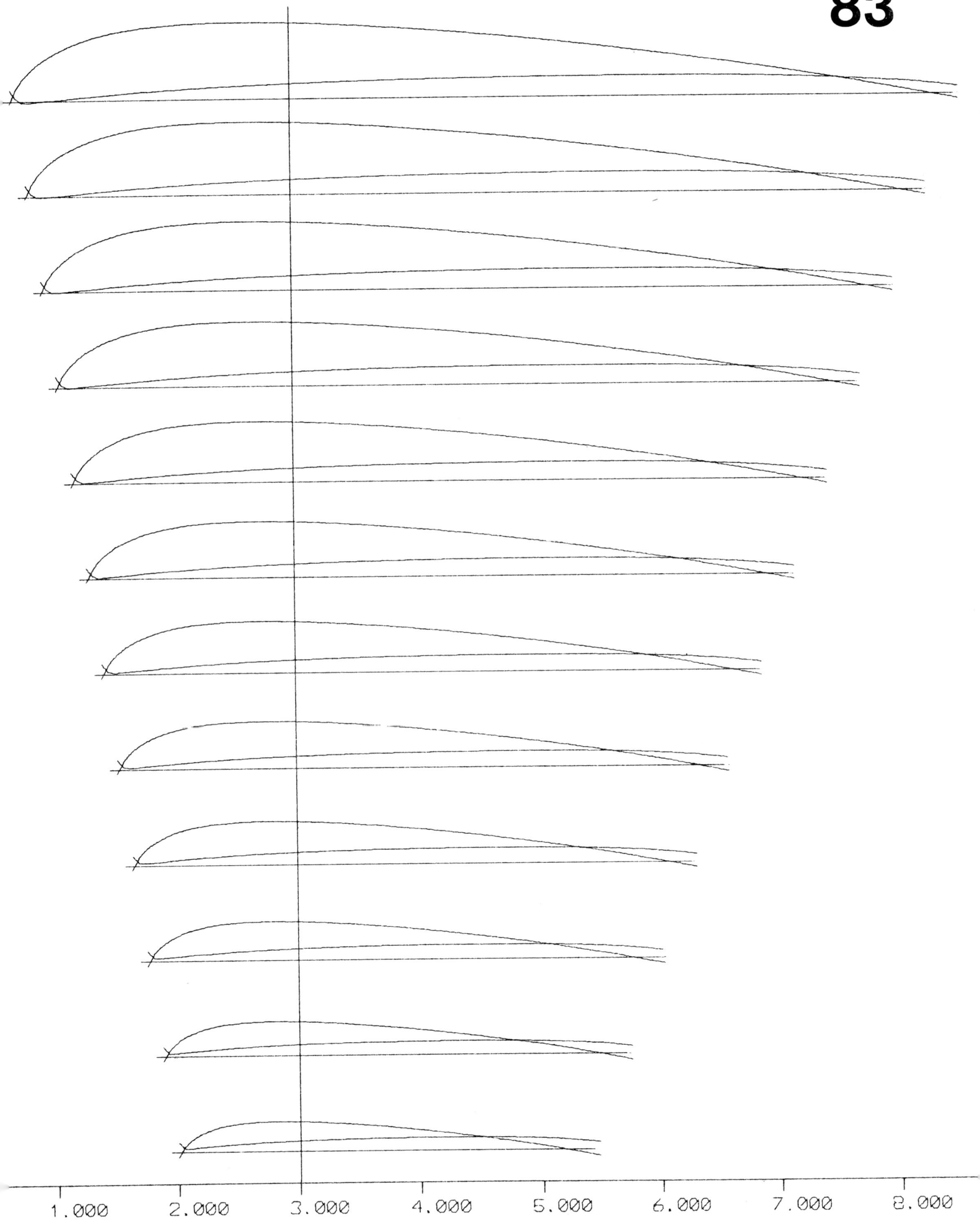

Index to Aerofoil Sections

Presentation

'Standard':	Full aerofoil outline. Nose requires rounding except sections marked * which remain sharp.
'Skimmed':	1·5 mm or 0·8 mm removed from top and bottom surfaces, to suit sheet covering or capping strip allowance.
'Blended':	Gradual change of section to eliminate bottom surface camber at wing tip.

Key to Model Types

F1A, F1B, F1C:	International Free-flight Classes, Glider (A2), Rubber (Wakefield), Power respectively
O/G, O/R, O/P:	Free-flight Open Classes, glider, rubber, power respectively
A1, C/D, $\frac{1}{2}$A:	Free-flight small classes, glider, rubber (Coupe D'Hiver) power, respectively
C/L, R/C:	Control line, radio controlled respectively

Aerofoil	Chord Range mm	Presentation	Model Type	Sheet
Benedek 6356B	200– 90	standard	F1A, A1	1
,, ,,	,,	skimmed 1·5 mm	F1A	2
,, 6407E	,,	standard	Jedelsky wings	3
,, 6456F	160– 90	,,	F1B	4
,, 7406F	,,	,,	F1B	5
,, 8356B3	200– 90	,,	F1A, O/G	6
,, 8535B2	,,	,,	F1C	7
,, ,,	,,	skimmed 1·5 mm	F1C	8
,, 8556B	,,	standard	F1A, O/G, A1	9
,, ,,	,,	skimmed 1·5 mm	F1A	10
Clark Y	200– 90	standard	R/C power	11
,,	,,	skimmed	,,	12
Czepa	200– 90	standard	solid wing F1A	13
Davis A.93, B.17	160– 90	standard	O/R, $\frac{1}{2}$A	14
Eppler 374	250–200	standard	R/C power	15
,, ,,	,,	skimmed 1·5 mm	,,	16
,, 385	,,	standard	R/C soarer	17
,, ,,	,,	skimmed 1·5 mm	,,	18
,, 385–387	250–180	blended	outer wing	19
,, ,,	,,	blended, skimmed	,,	20
,, 387	250–200	standard	R/C soarer	21
,, ,,	,,	skimmed 1·5 mm	,,	22

Aerofoil	Chord Range mm	Presentation	Model Type	Sheet
French	200– 90	standard	F1C, O/P, ½A	23
,,	,,	skimmed 1·5 mm	F1C	24
Göttingen 417	200– 90	standard	F1A, A1	25
,, 495	160– 90	,,	O/R, C/A	26
,, 602	200– 90	,,	F1C, O/P, ½A	27
,, 602	200– 90	skimmed 1·5 mm	F1C	28
,, 795	,,	standard	F1C	29
,, ,,	,,	skimmed 1·5 mm	,,	30
Hagel *	200– 90	standard	F1C, O/P, ½A	31
,,	,,	skimmed 1·5 mm	F1C	32
,, *	190–140	blended	outer wing	33
,,	,,	blended, skimmed	,, ,,	34
,, *	170–120	blended	,, ,,	35
,,	,,	blended, skimmed	,, ,,	36
Hansen AH–6–40–7	200– 90	standard	F1A, O/G	37
,, ,,	,,	skimmed 1·5 mm	F1A	38
,, ,,	150– 90	blended	outer wing	39
,, ,,	,,	blended, skimmed	,, ,,	40
,, ,,	140– 80	blended	,, ,,	41
,, ,,	,,	blended, skimmed	,, ,,	42
,, AH–7–47–6	200– 90	standard	F1A	43
,, ,,	,,	skimmed	,,	44
Hatschek	160– 90	standard	O/R, ½A	45
Horejsi	200– 90	standard	F1A	46
Isaacson 64009	160– 90	standard	O/R, ½A	47
Kaczanowski GF6	200– 90	standard	F1A, A1	48
,, ,,	,,	skimmed	F1A	49
Koster 66*	160– 90	standard	F1B, C/D	50
,, ,,	,,	skimmed 0·8 mm	F1B	51
,, ,,*	120– 80	blended	outer wing	52
,, ,,	,,	blended, skimmed	,, ,,	53
,, ,,	110– 70	blended	outer wing	54
,, ,,	,,	blended, skimmed	,, ,,	55
Lindner	200– 90	standard	F1A	56
MVA 301	250–200	standard	R/C Soarer, ½A	57
,, ,,	,,	skimmed 1·5 mm	,,	58
NACA 0006–0018	250	standard	R/C, C/L aerobatic	59
,, ,, ,,	,,	skimmed 1·5 mm	,,	60
,, 2312	250–200	standard	R/C power	61
,, ,,	,,	skimmed 1·5 mm	,,	62
,, 2412	,,	standard	,,	63
,, ,,	,,	skimmed 1·5 mm	,,	64
,, 4409	200– 90	standard	F1C, ½A	65
,, ,,	200– 90	skimmed 1·5 mm	F1C	66
,, 4412	250–200	standard	R/C power	67
,, ,,	,,	skimmed 1·5 mm	,,	68

Aerofoil	Chord Range mm	Presentation	Model Type	Sheet
NACA 6409	160– 90	standard	O/R, $\frac{1}{2}$A	69
,, ,,	250–200	standard	R/C soarer	70
,, ,,	,,	skimmed 1·5 mm	,,	71
,, 6409–2409	250–180	blended	outer wing	72
,, ,, ,,	,,	blended, skimmed	,, ,,	73
Neelmeyer	200– 90	standard	F1C	74
,,	,,	skimmed 1·5 mm	,,	75
Ritz 7·45–5·5G	200– 90	standard	F1A	76
Schwartzbach 68*	160– 90	standard	F1B, C/D	77
,, ,,	,,	skimmed 0·8 mm	F1B	78
Schweinsberg	160– 90	standard	F1B, C/D	79
,,	,,	skimmed 0·8 mm	F1B	80
Thomann F4	200– 90	standard	F1A	81
Wortmann M2	200– 90	standard	F1C, O/P, $\frac{1}{2}$A	82
,, ,,	,,	skimmed 1·5 mm	F1C	83